D1483214

Dedication

To Mom and Dad for your undying love and support.

CONTENTS

INTRODUCTION

If I were asked to describe building a battling robot in one word, that word would be *balance*. To build one of these things you have to balance your ideas and dreams with your time, ability, and money. You must balance speed with power. You must balance materials and weight limits. You must balance complexity with simplicity. In every aspect of the design and build process you must make trades between two or more desirable features. An old saying sums it up. "It's your choice…Fast, Cheap, or Reliable. Pick two." Balance is the key to building an effective, lasting robot. However, luck and driving skill have a lot to do with winning.

Building a battling robot can be a difficult, rewarding, educational, and really fun prospect. *Combat Robots Complete* was written to help you understand exactly what it takes to turn ideas into a functioning machine, and to answer a lot of the questions a person new to fighting robotics would ask without the burden of knowing all the physics, engineering, and math involved. I'm not saying there is no physics, engineering, or math in the book. I will attempt to present and explain a lot of the ideas behind building a battling robot, certain concepts are covered and explained in a simplified manner that may leave out details that would appear in engineering texts. Most of that isn't necessary to enjoy building and competing battling robots anyway. Like someone once said, "It ain't rocket science!"

No matter how hard you try, you cannot build an effective combat robot without having some mathematics skill. Calculus isn't required, but a good background in algebra, geometry, and unit conversion will get you a long way. I'm not writing a mathematics text, but throughout the book I will go through some basics in math and unit conversion that will help you figure out what kind of motors you need, what kind of batteries you need, what kind of gearing you need, and other important aspects of bot building. If you can substitute your values for the values given in the examples, you should do well.

Costs

Lots of people wonder how much money it takes to build a fighting robot. The answer is that it depends on the builder, his/her experience, his/her knowledge of the subject matter, and his/her wallet. When building a battling robot, there are lots of expenses. The best way to minimize them is to have a clearly defined plan and proceed according to it. Then you will most likely still be over budget. One major expense that most people forget to include in fighting a robot is the cost to compete. This cost includes the entry fee to the competition, hotel fees, rental car fees, airfare, robot shipping, meals, and any parts you may need but don't have for repairs during the competition. You can cut all that cost out, build a fighting robot, and stay home fighting the neighbor's trash can or chasing their cat—but that's not why you are reading this book. You want to compete with the big boys.

You can reduce costs in many ways. Booking flights with layovers will cut your airline costs. Planning a carpool with other builders for the event will save some money when it comes to traveling while competing. The local youth hostel may not be glamorous with its wards and cots, but it is a really cheap place to stay. Most of them even have kitchens. You can save some money by buying and preparing your own food. (You may put in some late nights getting your bot working, so you need to remember to check the kitchen hours and make other arrangements if necessary.)

Any way you look at it, shipping a robot through the standard carriers is expensive. The expense grows as you ship heavier and bigger boxes. Remember that you are not only paying to ship the weight of the bot. You are paying to ship all the spare parts, all the tools, and the crate. It will even cost you to build an appropriate crate. The first time I went to a major event I was on a really tight budget. We had a 300-pound bot that needed to get from North Carolina to Las Vegas, Nevada. The night before we left, we took the bot apart, packed it into seven cardboard boxes, and checked it all as luggage on the airplane. Four of us were going and that left only one available luggage slot. One of my crew was my friend's wife. I think she put everything she owned into that one suitcase and the rest of us took carry-on bags full of clothes. According to airline regulations, each box had to weigh less than 70 pounds. We were relieved to find out that the heavy one weighed in at 69.9 pounds.

I highly advise against shipping your bot as luggage for four reasons:

- Heavy boxes put the "lug" in "luggage." Getting to and from an airport is hectic enough without having to worry about loading and unloading all those boxes.
- You never know when an airline is going to lose your luggage. We got lucky and everything showed up in Vegas like it was supposed to but I know a few builders who have lost major parts this way.
- Airline screeners are much stricter about what they will allow inside the plane nowadays.
- It is a really bad thing to show up at an event with a robot that isn't put together. Doing so will take much of the enjoyment out of your experience. We put our 300-pound robot

together in the hotel room. It took about 20 hours—20 hours we could have spent talking to other builders and checking out the other robots.

The cost of competing is just another area where you need to compromise. You either spend some money or you spend some time. It is up to you to determine which is more valuable. Again, there are many ways to cut the costs. In fact, there are ways of building and competing for free, though it's not easy to get to that level. I'll cover those later.

Builder's Unions

In my experience, the best part of battling robots, next to building them, is getting to meet and talk to the other builders. I've made friends all over the world through robotics. Some I've met at competitions and some I've only talked to via Internet forums. The builders as a whole are a true community. This is great for a couple reasons. One, you get to share ideas and experiences. Two, you get to be a part of something that many people will only dream of doing. Everyone seems to stare in awe when I tell them that building robots has taken me to Hollywood, Las Vegas, and London, among other places. Whenever there is a big group of people, organizations tend to appear. The Society of Robotic Combat (S.O.R.C.) was the first of bot builder's unions, so to speak.

S.O.R.C. Mission Statement

First and foremost, the SORC is an organization devoted to the interests of robot builders from around the world. Where appropriate, the SORC will defend the rights of the builders to use their creations and images of their creations in any legal manner they choose. The SORC is a resource to provide education and information about the design and construction of competitive robots. Information about mechanical components, electrical components, and manufacturing techniques will be exchanged. Advances in Robotic technology will be identified and disseminated. The SORC will maintain a comprehensive set of guidelines for robotic events. These guidelines will maintain safety while stimulating creativity and impartiality.

Copied directly from the S.O.R.C. bylaws section 2.1... www.sorc.ws

To date, the major robot combat competitions have their rule sets based at least loosely on the S.O.R.C. rule sets and run their competitions according to safety guidelines originated by the S.O.R.C. The organization is also currently working on hotel and shipping discounts for builders who attend major competitions. The S.O.R.C. is not the only organization lending a helping hand to bot builders. Check on the Internet to find others. Become and active member and help make this sport grow.

Following the Rules

Many people start out imagining robots carrying flame-throwers, lasers, and even firearms. In reality, most competitions don't allow these types of weapons. If they are allowed at all, they are so restricted that they could not possibly do any damage to their opponent. One of the first things you have to do before building your bot is to know the rules of the competition in which you wish to participate. Notice I said "know" the rules and not "read" the rules. Knowing the rules will keep you from making design mistakes and will ensure that you are allowed to compete once your bot is finished. Knowing the rules will allow you to design for that specific competition as well. Some competitions have different events with different strategies. Simply reading the rules just doesn't cut it. Knowing the rules of the game will give you an edge compared to the builder who does not know them.

While attending an unnamed competition that required the breaking of a lot of glass, I noticed that my robot would not be able to reach the glass. Several other builders had the same problem. Evidently I did not pay close enough attention to the rules. Fortunately, we had some spare material that did the job nicely. However, some of those other builders did not have the spare stuff or a place to mount it. They ended up losing points in the competition. That could have been avoided if they knew the rules instead of simply reading them. I just got lucky because I brought a ton of stuff.

Weight Classes

The most common rule between different competitions is the weight class rule. For all the nitpickers, we should remember that weight is different than mass. Mass is constant where weight varies with different forces. The weight of a robot hovering above the scale is zero because of the forces holding it up in the air. However, the mass of the robot is the same whether it is in the air or on the ground.

Among the competitions that I know of, there are seven weight classes ranging from a few ounces up to 408 pounds. Some classes contain exactly the same weight ranges, while some classes are just similar enough to facilitate losing a few pounds to be eligible to compete in different competitions. Most competitions offer a weight advantage to robots that walk. The spirit of the rule grants the weight advantage because it is recognized that it takes a bit more to build a walking bot that can compete effectively with a rolling bot. For now, the rules dictating what exactly constitutes a walker are really still under debate. Some rules are too restrictive, and others just are not restrictive enough. Because of that and the added complexity of walking machines, only the most adventurous bot builder attempts the walker undertaking. If you are reading this book in order to start building your first bot, I highly recommend starting with a wheeled bot so that you can learn to walk before trying to run.

Very few competitions allow all seven classes. For instance, the Ant weight class is from 0 to 1 pound for rolling robots. While they are a great pastime, those robots are just too small to put into a 30-foot square arena and entertain a crowd. For now, the Ant weight class has been relegated to separate, side competitions, while the larger bots are in the big arena.

One thing to remember when it comes to a real competition is that the weight class rule should be set in stone. At least one of the TV shows will allow you to compete even if your robot is overweight. The real competitions will not do this unless you change official weight classes. However, the real competitions will allow you to try and drop the extra weight to get into the intended class.

Technical Concepts

In this book, I'll outline and discuss the major concepts behind building a combat robot. The information is not new to machinists, mechanics, and electricians, but it is finally set down in a manner that is appropriate to robot builders. I'll start out with the schematics, wiring, remote control setups, and their problems and uses. Then I'll touch on different methods of movement. Next I'll cover the popular speed controllers. Then I'll cover batteries, including SLAs and NiCds and how to charge them. After that, I'll go into electric motors and show you how to choose the right one. I'll also go into how to increase their torque for your bot using gears, sprockets, or pulleys. I will touch on pneumatic systems, getting the bot to roll smoothly, and cover the use of gasoline engines. Then we'll come to frame and armor material selection. We'll discuss the different types of weapons and defenses. Finally, we will talk about how to design your bot and model it so that we know it works. That will do it for the conceptual part of the book. After that, I will briefly discuss building a 1-pound robot and a 220-pound robot. The third and final project of the book will be Dagoth, a 30-pound bot armed with spikes and a wedge. It will not be a large stretch of the imagination to turn Dagoth into a 60-pound bot with an active weapon. Have fun.

COMBAT ROBOTS COMPLETE

1

ELECTRONICS AND WIRING

This section of the book is in no way expected to teach you everything about electronics and wiring. However, I have included enough theory to explain what you need to know to get started in combat robotics.

The *atom* is the heart of electronics. Atoms have a center, or *nucleus,* made up of *protons* and *neutrons*. Think of the cluster of protons (positive charge) and neutrons (no charge) as a planet. Atoms also have *electrons* (negative charge) that circle around the nucleus like moons around the planet. The atoms can have a surplus of electrons, a shortage of electrons, or just the right number of electrons. Atoms that have a surplus can easily combine with atoms that have a shortage. They share the electrons and in the process create compounds. Compounds can still have a surplus, a shortage, or just the right number of electrons. The compounds that make up wood or glass have the right number of electrons. This makes them *insulators*. Compounds with a surplus or a shortage are called conductors. (We don't need to discuss semiconductors here.) Copper and steel are examples of conductors. The atoms that make up copper and steel share electrons. It is this sharing that makes a *circuit* possible.

Batteries have a positive and a negative *terminal*. The negative terminal has lots of electrons. The positive terminal does not. Everyone knows that opposites attract. The electrons want to get to the positive side of the battery but cannot unless a circuit is made. *Current*, or electrons, flows from the negative terminal of the battery, through the *load* (motor, light, etc.), and back into the battery through the positive terminal by sharing electrons through all the conductors. This is called a circuit. Once most of the electrons have traveled from the negative side to the positive side, the battery is dead. If there is a break in the path of the current at any point, including inside the motor, the current flow stops and the motor stops spinning. If a part of that circuit takes a shortcut, say from one battery terminal to the other without going through the motor, the motor stops but the current continues. This is called a *short circuit*. This is a really bad thing to have happen. It can damage your batteries, because there is nothing there to regulate the flow of electrons.

That is about the extent of the electronics philosophy required to apply what you learn from this book to robot building. However, I do explain the techniques I use to illustrate the wiring diagrams presented. The *schematic* diagrams are not exact wiring schematics with respect to certain speed controllers or remote control interfaces. They are block diagrams that show you the basic layout for our intended uses. In most cases, I also include a picture or drawing that shows the actual component and wiring. Refer to an individual device's owner's manual for exact placement of wires. Check the "Electronics" section of Appendix E for some great books on getting started with electronics. At the end of this chapter, I go step-by-step through a few simple example circuits that are really important to bot builders.

Schematics

You must be able to recognize certain objects in the wiring diagrams to understand how the schematic should work. There are twenty-two lettered objects in Figure 1.1 that correspond to the twenty-two lettered objects in Table 1.1.

TABLE 1.1 SCHEMATIC FIGURE DESIGNATIONS AND DESCRIPTIONS

FIG.	NAME	DESCRIPTION
A	Wire	Connects different components. Thicker wire can handle more current. Thick lines indicate that you should use heavy-gauge wire.
B	Connection	Wire connections are indicated in this book with a dot.
C	No Connect	A non-connection between two wires that cross is indicated without a dot.
D	Terminal	An input or output of a device.
E	Polarized Terminal	An input or output of a device that should be positive or negative. This is indicated by the plus or minus sign next to the terminal.
F	Supply Voltage	Indicates that this line is connected directly to a positive terminal of a battery.
G	Ground	Indicates that this line is connected directly to a negative or ground terminal of a battery.

continued on next page

TABLE 1.1 SCHEMATIC FIGURE DESIGNATIONS AND DESCRIPTIONS (continued)

FIG.	NAME	DESCRIPTION
H	Battery	Shows the positive and negative terminals of the battery. Will also show the voltage of the battery. Sometimes one battery with a higher voltage number will be used to indicate what is really one or more lower voltage batteries wired in series.
I	Switch	A device to make or break a circuit. Includes all manner of switches. Shown is a single-pole single throw switch (I-a) and a double-pole single-throw switch (I-b).
J	Coil	A winding of wire used to create a magnetic field. Coils, as referred to in this book, are used in relays and contactors or as solenoids to electromagnectically actuate the switch(es) inside the relay, contactor, or solenoid.
K	Contactor or Solenoid	An electromechanical switching device for controlling a high voltage or current with a low voltage or current. Usually has a single electromechanical switch and is normally rated to handle higher currents.
L	Relay	An electromechanical switching device used for controlling a high voltage or current with a low voltage or current. Can contain many electromechanically actuated switches. Not usually rated to handle higher currents, but can be used to actuate a contactor or solenoid.
M	MOSFET	A solid-state electrical switch used in speed controllers.
N	Capacitor	Used to limit motor noise. An electricity storage tank.
O	Diode	Used to limit relay, contactor, and solenoid generated spikes. Allows electricity to flow in one direction only.
P	Resistor	Used to limit current. Sometimes used in conjunction with a meter to calculate current (in that case it is called a shunt).
Q	Motor	Drive or weapon mechanisms. Could be used in place of the actuator icon. Reverses spin direction when the polarity of the applied voltage is switched.
R	Actuator	Drive or weapon mechanisms. Should not be used in place of a motor icon.
S	Meter	Used to measure many electrically related values, including voltage and amperage. The arrowhead leads indicate test connections.
T	ECL	Emergency cutoff loop. Kills power to the entire robot.
U	Antenna	Helps transmit and receive remote control signals.
V	Any Device	Usually a block with labeled terminals. Takes the place of anything that makes the schematic too complicated to draw. Also takes the place of standardized devices where it would not be prudent to include the exact wiring diagram or for which you do not have the exact schematic.

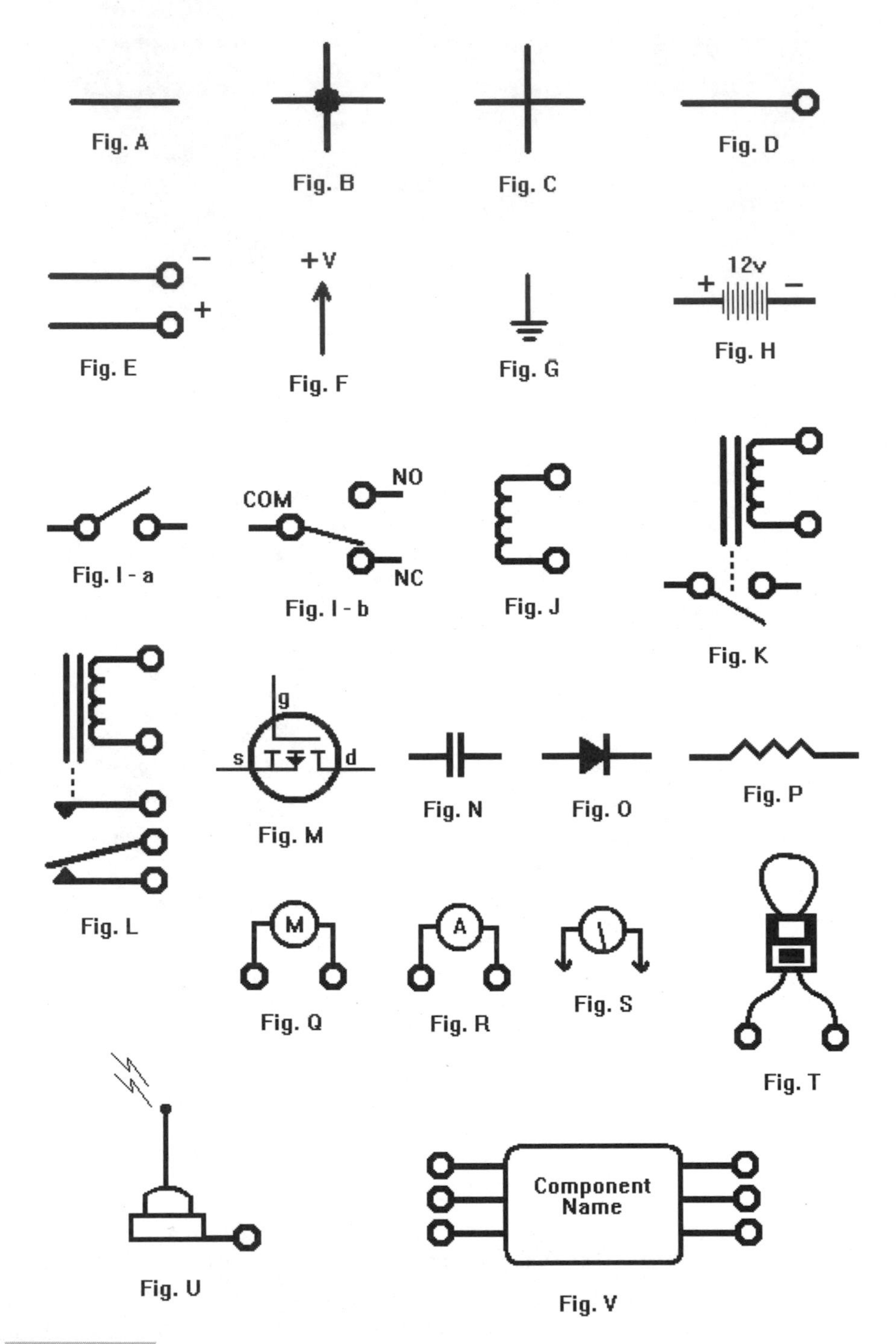

FIGURE 1.1 Electrical icons used in this book.

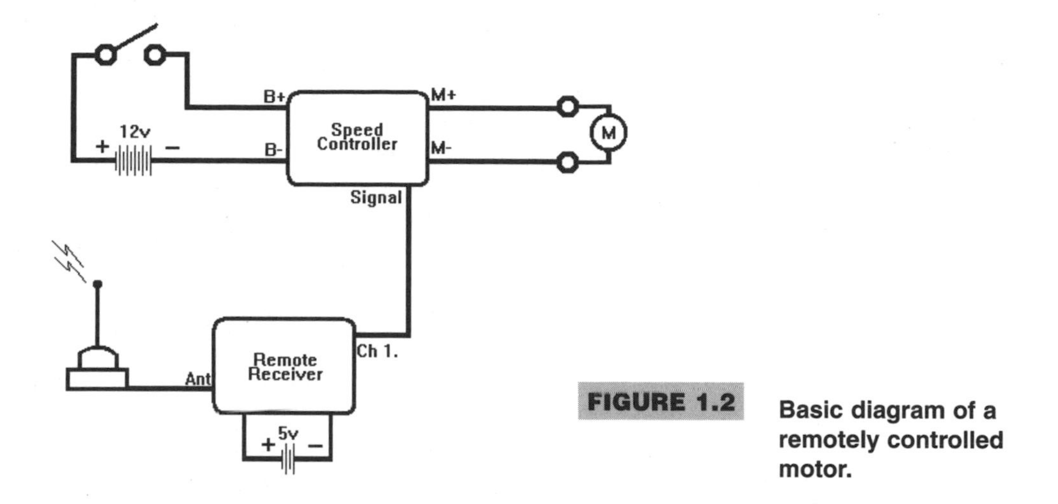

FIGURE 1.2 **Basic diagram of a remotely controlled motor.**

Now that you know the icons used in the book, Figure 1.2 will get you used to seeing them in action. It clearly shows a motor (M) connected to a *speed controller*. The speed controller has inputs for motor power (B+ and B−) that are connected in series with a battery and a switch. The speed controller also connects to the remote control receiver through a Signal wire. The *remote control receiver* connects to the antenna and is powered by a 5-volt battery. This is not an exact diagram of a safely working mechanism, because a couple of components are missing, which we will get to shortly.

Wire Size

One of the most important things in wiring your robot is the selection of the correct wire size. Small wires handle small amounts of current. Large wires handle large amounts of current. If you try and run too much current through a small wire, the wire heats up and eventually burns the insulation. This can cause shorts in overlapping wires or open circuits if the wire itself burns in two. Appendix C contains a "Chart of Wire Size, Current and Length" that shows the amounts of current appropriate for each size and length of wire. Appendix C also includes a chart of the "American Wire Gauge (AWG)/Diameter/Resistance" showing the sizes and resistance of each wire. Note that wire size goes down as *gauge* (AWG) number goes up.

Wire sizing is a tricky business. Even though I offer a chart, finding the correct amount of current in your system at any one time is the key to the size of wire you should use. If you are using motors that draw 150 amps at stall (see *stall current* in the glossary), you should not have to use the 4 AWG wire, about as big as a small finger, to wire your bot. Designing the *drivetrain* correctly ensures that you never see the total 150-amp stall condition unless something goes seriously wrong. In that case, wiring is not your major worry. (I cover drivetrain design in Chapters 6, 7, 8, and 21.)

Let's continue with the example at hand for now. That same 150-amp stall motor may only demand 70 amps at any one time, depending on how it is geared. Because of that, you should be able to get away with using 8 or even 10 AWG wiring. (I cover how to figure out how much current your motors are likely to pull based on gearing in Chapter 6.) Factory wire size and component ratings do not necessarily agree with each other. Many times you will have a motor that is rated to pull 200 amps in a stall condition, but the wires attached to it are 10 AWG from the factory. This is because the motor is not designed to run at stall. Factory conditions do not usually put the kind of loading on a motor that it will see in a robot battle. When you come across a motor that is wired like this, and you know that you will be drawing more current than the wires can handle, you have only two choices. You can find a new motor or attempt to install larger wires. I suppose you can leave it alone and let the wires burn up, but why?

One other aspect of wire sizing is the amount of *voltage drop* you will experience. Several things cause voltage drop. In some cases, the battery cannot source the amount of current your motors want. When that happens, the actual voltage coming from a 12-volt battery can drop way down. In that case, you need more batteries to help source the current demanded. A battery that has gone bad in some way can also introduce voltage drop, if it still works at all. In either case, depending on how your speed controllers are wired, this can cause a system shutdown.

The length of wires can cause voltage drop, too. Wire is nothing more than a resistor with an extremely low resistance. However, the resistance exists. Using Ohm's law (Voltage = Current × Resistance, or $V = IR$), the given current and the resistance of the length of wire, you can figure out how much of a voltage drop you will experience. Use the chart of "American Wire Gauge (AWG)/Diameter/ Resistance" in Appendix C to find the resistance. Multiply that by the amount of current you will be using. The answer is the amount of voltage drop. The numbers are small until you start using really strong motors that draw a lot of current. It just so happens that this is when you really need to worry about it. If you keep your wires as short as possible, you will not experience this problem drastically. However, short wires make it more difficult to work on and repair your robot. Voltage drop is something to keep in mind when you are designing that 200-pound spinner robot. Voltage can drop significantly because of the battery type, current drawn, and wire size and length.

Connectors

When wiring motors, batteries, speed controllers, and whatever else you might have, you do not want to twist the bare ends of the wires together and wrap some electrical tape around them to make your connection. You do not want to use wire nuts either. Either of these methods is fine for home wiring, but your robot moves and will take substantial shock loads. These methods are unprofessional, unreliable, and unsafe in fighting robots. *Terminal connectors* are the way to go. There are literally hundreds of types of connectors you can buy. They are usually color coded according to the amount of current they can handle or wire size they can handle. You do not need to memorize the color-coding scheme to use terminals. Terminals usually fit nicely on the wires for which they are intended. The sizes are also indicated on the packaging. Having the correct wire rating is what is important. If you are trying too hard to push a connector onto a wire, you should get a larger connector. If the connector dwarfs the wire, you need a smaller connector. Using the wrong size of connector causes bad or loose connections in your system.

FIGURE 1.3 **Ring terminals and PowerPole brand connectors.**

To use connectors, you will need a special tool called a *crimper*. A crimper squeezes the terminal end onto the wire and creates a strong mechanical link. Some terminals require a special crimper that can cost a lot of money. It is my opinion that you can use the more common connectors and their cheaper crimpers with success. Of the major types of connectors, I usually have a large stock of ring terminals and some end connectors, as seen in Figure 1.3. Ring terminals are used where there is a screw that will secure the terminal to its connection. I use ring terminals for one reason: if the screw vibrates loose, you still have a connection. If you use the fork-type connector, it can fall out when the screw loosens. The end connectors I like to use, Anderson PowerPole brand, can be disconnected easily but also have a positive locking force to keep them together.

Things to Remember

Wires come in different colors for a reason other than because they are pretty. Wires are color-coded. It seems there isn't one recognized coding standard across industries: the telephone industry is different from the automobile industry, and both of those are different from the airplane industry, which is different from the computer networking industry. Even the servos that come with model planes have different wiring colors and pin placements among manufacturers. A "Chart of Servo Lead Colors by Manufacturer" is given in Appendix C. It lists the power lead, ground lead, and signal lead colors of the major RC manufacturers. Depending on which brand of remote control you buy, you will need to know which pins are what on the receiver so that you can connect any third-party devices meant to control your bot.

The wiring of the receiver is not the only wiring that should be color-coded. All speed controllers that you buy commercially have positive and negative terminals for the motor power supply. Sometimes there will be a sticker with the positive and negative designations on it, sometimes there will be small plus and minus signs next to the terminals, and other times the terminal housing will be red or black. Red is for positive and black is for negative. Use this color code to keep the wiring inside your robot simple, too.

Wires can get really messy. Use the shortest lengths possible to reduce noise, but you also need some room to make changes and be able to work on the bot. I have seen robots that looked like they had a bird nesting right in their middle. This is no way to wire a bot, because it is next to impossible to troubleshoot anything in that condition within a time limit. I like to use *wire ties* and *sticky backs* (see Figure 1.4) to secure my wiring out of the way and in a neat fashion. It makes the job look more professional and makes it easier to work with. Many people go as far as running conduit or some other type of wiring path. Although it is a bit of extra work and is not totally necessary, this practice sure makes the robot look nice.

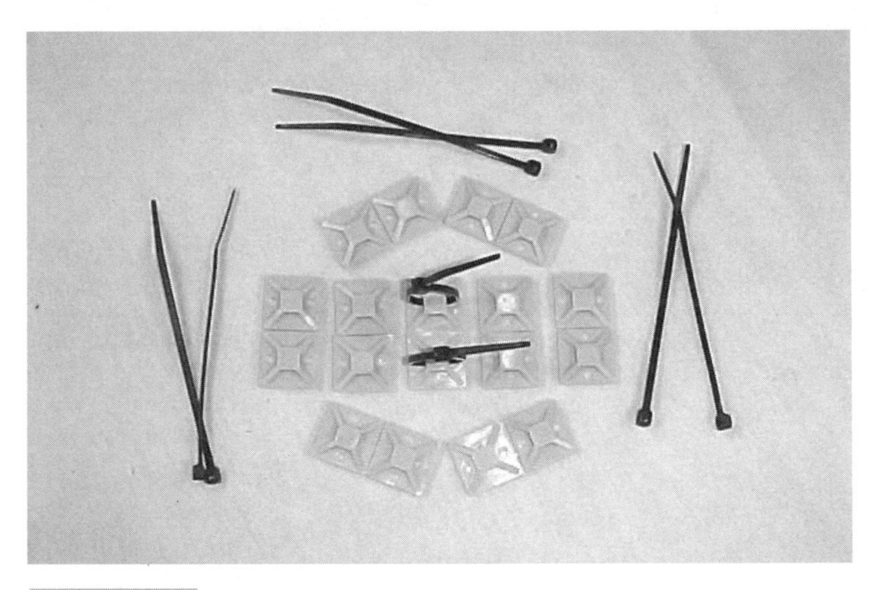

FIGURE 1.4 **Wire ties and sticky backs.**

Shutting Down

This is as good a time as any to discuss how to turn the power to your bot on and off. Most competition rules have several things in common. The main power cutoff rule is one of those. Preferred methods of cutoff may differ, but the rule is still there. The most common requirements state that you need an *emergency cutoff loop* and a *main power switch*. At first glance that may seem redundant, but both are truly necessary in the dangerous robots. Some builders go one step further and include a way to kill the main power remotely.

EMERGENCY CUTOFF LOOP

The *emergency cutoff loop* (ECL) is in the rules for two reasons. The first, obviously, is so that there is a fast, simple method for killing the power to the robot. The second is so that other people can positively know that your robot is powered down just by looking at it. If the loop is disconnected and lying on top of the bot, it cannot be powered up. The ECL is kind of a deceptive name. In most situations you will not approach an out-of-control robot, but if you do, you should simply let the batteries run down. However, if the robot is causing harm to a person, you should make every attempt to disconnect the loop.

The ECL is simply a wire loop with a connector on it. Both the wire loop and the connector should be sized so that they can handle the entire current demand of the robot. The connector plugs into the opposite gender type of the same connector mounted securely on your robot. Notice in Figure 1.5 that the ECL is connected between one motor wire and the battery. The other motor wire goes directly to the battery. No matter what you do, the motor will not run if the loop side of the ECL is not installed. Usually the ECL is left disconnected until right before a match or during testing. For simplicity, this example does not include a speed controller or a remote-controlled contactor. Those diagrams are covered in the Simple Electronics Examples section at the end of this chapter.

SHUTDOWN PROCEDURE

When it comes to using remote controls to operate a dangerous machine with who knows what type of weapons, there is a procedure you need to follow in order to be safe. Depending on the type, radio receivers can take several seconds to settle down and send out clean, safe signals to your bot. If you do not start up correctly, the receiver can send unwanted signals to any part of your bot, including the weapons. Also, speed controllers and relays can do some strange things when powered up—even if the receiver is not on. The best way to power up your bot is follow these steps:

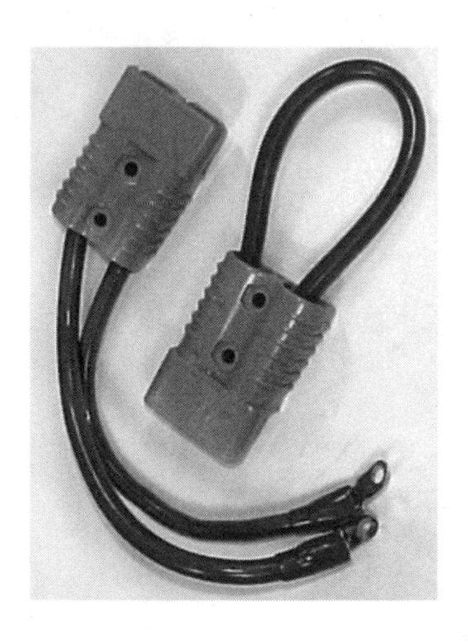

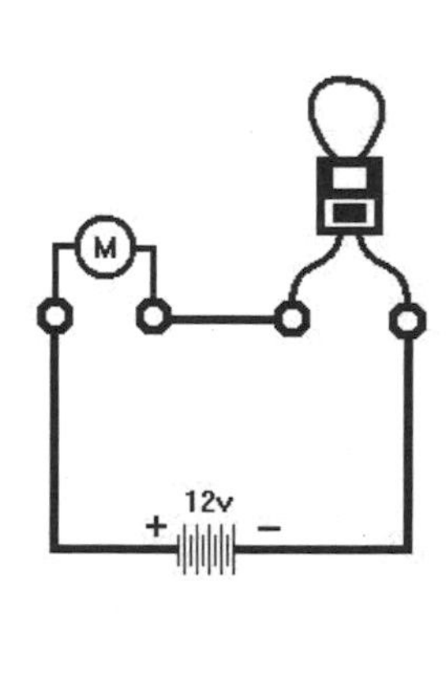

FIGURE 1.5 Picture and schematic of emergency cutoff loop.

1. Turn on the transmitter.
2. Make sure all the trim settings are correct and joysticks or wheels are centered.
3. Make sure all the switches are set so that weapons are deactivated.
4. Turn on the receiver and wait several seconds for it to settle.
5. Turn on the main power to the robot.

Never get close to the bot after it is powered up. Any kind of stray signal could possibly command it to lurch one way or another or to start spinning that 28-inch sawblade while your hand is resting on it. Deactivate your robot in reverse order. Use the emergency cutoff loop to kill the main power first. It should be visible and easily removed.

MAIN POWER SWITCH

The *main power switch* (MPS) is mounted in series with the ECL and should be large enough to handle the total current requirements of the robot. There are several large switches that are excellent for this type of job. I tend to use the Hella emergency shutdown switches from Flaming River, which are used on drag cars and boats. They can handle 200 to 300 continuous amps with 500-amp spikes. This is usually more than the robot will demand. For smaller robots you can use a switch similar to that shown in Figure 1.6. If you get something smaller, you could burn it up pretty quickly.

REMOTE CUTOFF CONTACTOR

Some builders go the extra mile and install a large *relay* or *contactor* that controls the main power along with the ECL and MPS. They connect the relay to a small device that turns it on or off depending on what signals it gets from the remote receiver. Team Delta sells these electronic decoders for the receiver for relay control. Some competitions even require this

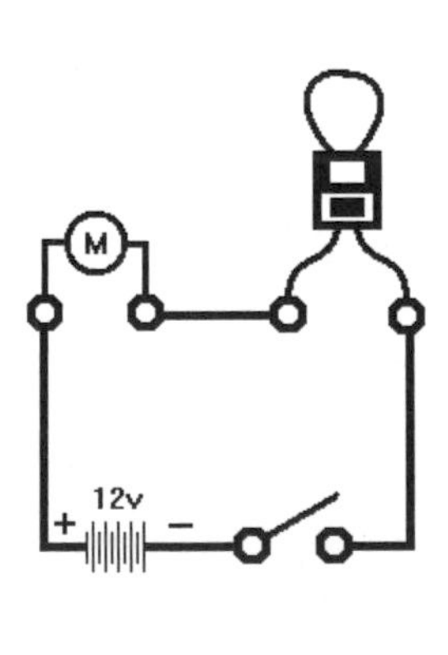

FIGURE 1.6 **Picture and schematic of ECL and master power switch.**

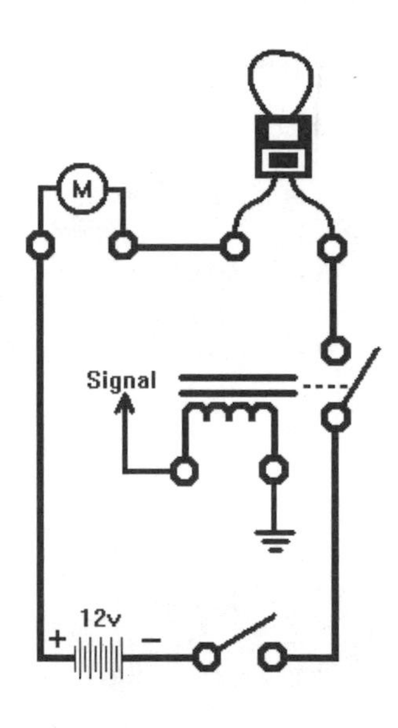

FIGURE 1.7 **ECL, MPS, and remote cutoff contactor.**

setup. To do this safely, you must make sure the main power will go through only the relay when it is energized. That way, the bot has no chance of being powered up while you are not in control. White-Rogers sells some nice, big contactors suitable for this purpose. I have also used the large contactors shown in Figure 1.7. They came from my local car stereo supply store. Calculate the total current that the contactor needs to switch before buying one to match.

Simple Electronics Examples

Many people who are new to electronics get confused when it comes to wiring a robot. There is no need for this confusion. Wiring a fighting robot is really simple. There are only a few types of circuits you need to worry about when trying to get your first robot running. I just covered the ECL and MPS. Shown earlier was an example of connecting a motor to a speed controller and battery. The following circuits are meant to help you get to know what is needed. If you are new to electronics, you might want to get the parts and build the small circuits just to get a feel for what you will be doing. You do not need to get a 1-horsepower motor and some 13-amp hour batteries to build them. You can use almost any small, direct current motor that runs on low voltage. The motor out of a cheap remote control car toy will work with a 9-volt battery. You can also buy small motors at your local Radio Shack. The concepts presented using the small components work exactly the same way for the bigger, more powerful components of your fighting robot. However, if you decide to wire up your powerful motor for these tests, please put it securely in a *bench vise*. Large motors tend to jump violently when full power is applied in the manner shown.

MOTOR SPIN DIRECTION

First, we will see how to reverse the spin direction of a motor. Being able to reverse a motor is critical. Motors control the direction of your robot. If you cannot run in reverse, what's the point? Changing the direction of current flow from the battery through the motor does the reversing. Simply put, you switch the wires around. That is one of the functions of a speed controller. Figures 1.8 and 1.9 show a motor and a battery. In Figure 1.8 the motor's positive (red or +) lead is connected to the battery's positive lead. The motor's negative (black or −) lead is connected to the battery's negative lead. The motor spins in one direction, forward.

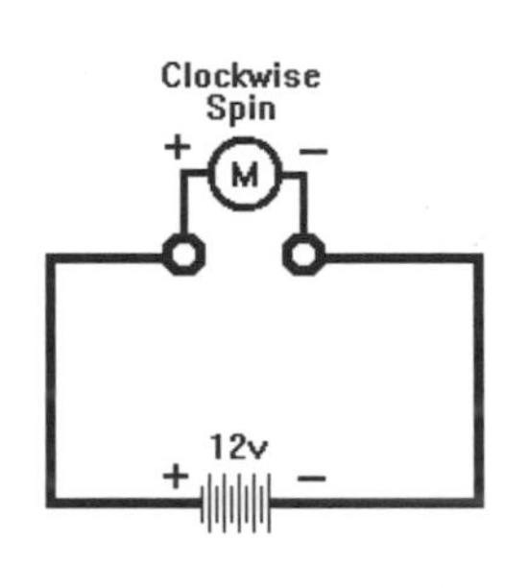

FIGURE 1.8 Forward motor spin. Lead colors match.

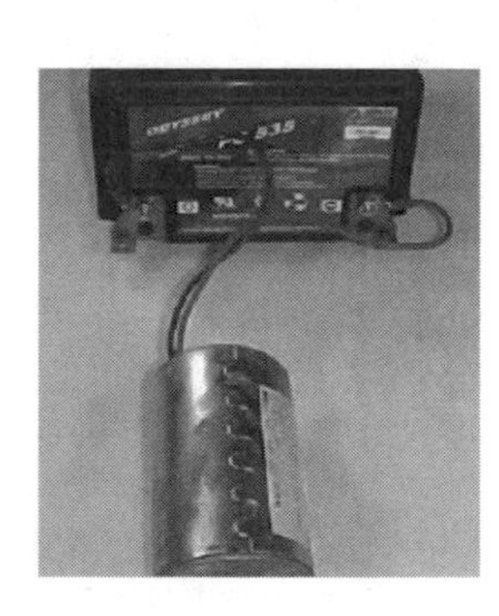

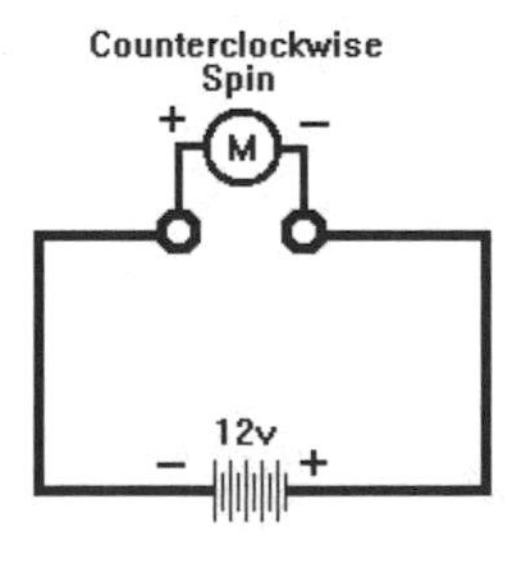

FIGURE 1.9 Reverse motor spin. Lead colors do not match.

Figure 1.9 shows the motor with the leads switched so that the motor positive touches the battery negative and the motor negative touches the battery positive. The motor spins in the opposite direction, reverse. You may want to put a wire tie around the shaft in order to easily see the direction of spin.

SWITCHES

Next, you wire a switch into the circuit. Switches can have two to ten or more terminals. The more terminals a switch has, the more contacts there are inside it. A *single-pole double-throw switch* (SPDT) (Part I-b in Table 1.1), it has three terminals. The center terminal is usually the common. If you actuate the switch in one direction, the center terminal and one of the outside terminals are connected. If you actuate the switch in the other direction, the center terminal and the other outside terminal are connected. The center terminal is the *common* reference point. Some switches stay connected only as long as you hold them closed. These are called *momentary switches* and are most commonly seen as push buttons. Momentary switches are pretty much the same as regular switches. Their terminals are usually marked with an NO, an NC, and a C or COM. NO means normally open. NC means normally closed. You should be able to guess that the C or COM means common. *Normally open* means that anything connected to that terminal will not be on unless you have the switch pushed in. *Normally closed* means that anything connected to that terminal will be turned off when you push the button and on when it is not pushed. Which terminal you use depends on what you want the circuit to do.

In this example, it does not matter which way you wire the motor and battery terminals, either positive to positive or negative to positive. Connect one motor wire to the battery positive. Next connect the battery negative to the common switch terminal. Connect the other motor wire to a different switch terminal. Actuating the switch turns the motor on and off. The spin direction does not change.

RELAY CONTROL

A relay is a switch that can be actuated with electricity. This is necessary when you are not able to actuate the switch yourself. A computer or some other electrical device can actuate the switch when needed. Relays have several terminals or pins. Each one is usually lettered or numbered depending on the manufacturer. Two pins are used to energize a coil that magnetically pulls the switches closed inside the relay. The other pins are connections to the switch(es). The manufacturer usually prints the schematic of a relay on the back of its packaging or on an attached piece of paper. Occasionally they print it directly on the relay casing. This schematic shows which pins are for the switches and which are for the coil.

Just as in the motor example in Figure 1.8, connect one of the coil terminals to one side of the battery. Then connect the other coil terminal to the other side of the battery. You should hear a click. Do not leave the coil terminals connected. You might forget about them and drain the battery.

A relay is not much use if you do not connect something to its switch terminals. Follow the diagram on the back of the relay package and determine which pin is the switch common. Connect one motor lead to that pin. Now find the normally open pin for the relay. If there are more than three switch terminals on the relay, make sure you find the normally open pin associated with the common pin you are using. Connect one of the battery leads to the normally open pin. Connect the remaining motor lead to the remaining battery lead. The motor should not be running. Now repeat the process of hooking each coil pin to the battery leads to turn on the relay. Use the second battery for the coil pins. When the relay clicks on, the motor should start to spin. In this example, you are using separate power supplies to run the motor and turn on the relay. In actual bot use, you would probably connect a higher voltage source to the relay switch pins to power the motor than you would use to turn on the relay coil.

DOING IT REMOTELY

You will probably use a small electronics board, like one from Team Delta, to control the relay using 5 volts. However, a small relay will burn up when you use it to control a motor suitable for your robot. To remedy this, you can buy a big relay or big contactor that is designed to handle the amount of current your motor wants. The only problem is that the big relays and contactors draw a good amount of current, and require more than 5 volts to switch on the coil. This could burn up the small electronics board you will use to operate it. The solution is simple. Use the electronics board to activate the small relay. Use the small relay to activate the large contactor. Use the large contactor to switch on the motor. In Figure 1.10, the RC interface runs on 5 volts supplied by the RC receiver. The RC interface must activate when 5 volts are supplied, but it can switch both more current and the 12 volts needed to activate the large contactor. The large contactor activates when 12 volts are supplied by the RC interface and can switch the high current and 24 volts that supplies the motors.

TAKING CARE OF SPIKES

Something you may notice is the addition of a *diode* (part O in Table 1.1) to the relay coil leads. Diodes allow current to flow in one direction only. In this example, the diode is connected so that it conducts current only when the relay is supposed to be off; otherwise it

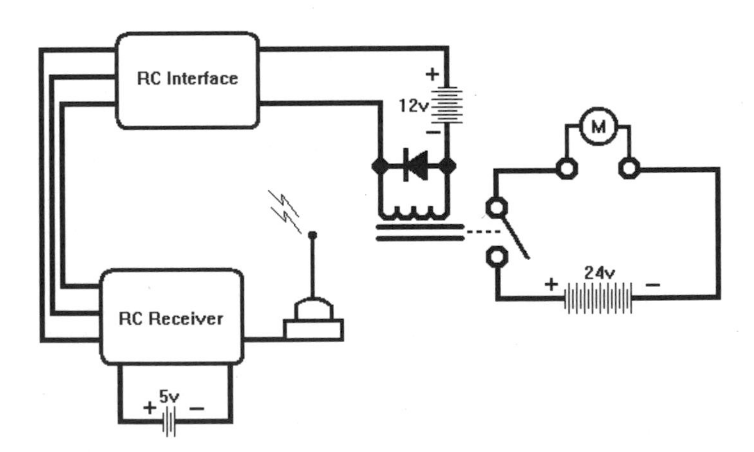

FIGURE 1.10 Using 5v to power an RX, 12v to power a relay, and 24v to power the motor.

would create a short circuit and the relay coil would never turn on. The reason for the diode is that the coil of the relay generates an electrical spike when power is removed. The spike is in reverse polarity with respect to the electricity that originally powered the coil. Because of the reverse polarity, the diode conducts the spike. The spike is then dissipated back into the coil instead of back into the electronics of the RC interface. Diodes for use in this manner are usually black with a white band around one end showing the direction of current flow (from positive to negative, white side is negative). The straight line in the diode's schematic symbol represents the white band. You can find diodes at any number of places, including Radio Shack. Make sure they are rated for at least three times the operating voltage.

GOING FURTHER

Only two things are missing from the last example. If you add an emergency cutoff loop and a master power switch, you will have the entire electrical system for any number of electrically powered robot weapons. Spinners use big motors that need to be turned on and off. Lifting arms can use motors. Swinging hammer weapons can use motors. If you want lights, replace the motor with a light and it works the same way. Anything that you want to turn on or actuate electrically can be controlled remotely with this setup. Pneumatic valves are electrically switched in the same manner. Remember to figure out the amount of current you will need to power each stage of the circuit, and then buy parts that will handle that current. In the previous schematics I used separate batteries to power separate stages of the control circuit, but that is not necessary. In the "Multiple Batteries" section of Chapter 5, I describe how to get different voltages out of multiple batteries wired in series. I also explain series and parallel wiring.

Summary

In this chapter, we discovered electricity, the life-blood of a robot. With technology in its present stage, the control systems of all but the most complex combat robots are simple on–off circuits coupled with plug-and-play electrical components. The simple electronic examples I have offered will get you started. Throughout the rest of the book, you will apply combinations of these circuits to form complete robot control systems.

In Chapter 2, we discuss the actual remote control system. We cover the different types of standard remote controller, and I will show you some simple modifications you can make to increase their usability and your comfort. FCC regulations play a small part as well.

2

REMOTE CONTROLLERS

One of the most important things in your combat bot is the *remote control unit*. This is how you control your bot. Several types of remote control are available. Starting with the popular radio bands, you have AM, FM, and digital. AM radio controllers are not suitable for combat robotics, because their susceptibility to interference is too high. Most serious competitions require you to operate on a FM radio controller at the very least. AM is just too noisy to control a robot that could do serious damage to people or equipment. AM radio controllers come with the less expensive remote-controlled cars and trucks.

PPM and PCM Radios

I've had my share of interference and lost matches while using an FM radio controller. FM radio controllers are normally split into two styles: *PPM-style* is standard and is the least expensive; *PCM-style* FM radios are a step up because they have a computer inside that codes each packet of information sent to the receiver. The receiver listens for this packet of information. When the receiver gets the information, its computer decodes the signal. If the transmission is garbled or has come from an outside source, the information is tossed out. Another plus about a computerized radio is the *failsafe* mechanism. These radios can be set to move their servos, in our case motors or actuators, back to a safe setting in case the signal is lost or there is too much interference. The failsafe settings are important to you and everyone who will be near whenever you operate your bot. They are well worth

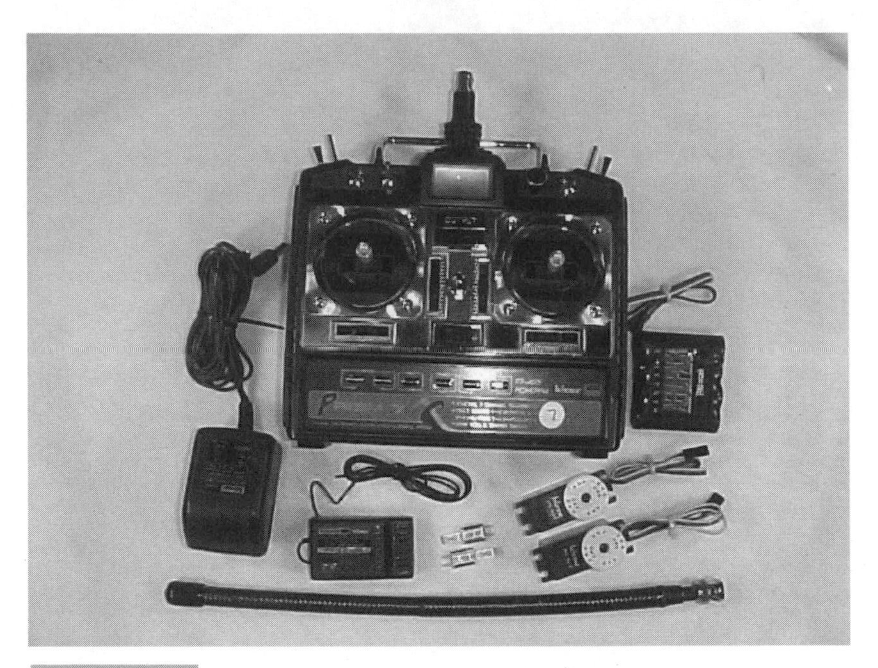

FIGURE 2.1 **Hitec Prism 7 channel PCM radio set.**

the extra money you spend to get a PCM-style radio, as shown in Figure 2.1. Consult your RC rig's manual on how to program them. If you don't have or plan to get a computerized radio, there are individual channel failsafe mechanisms available at most hobby stores, either online or down the street. Also, all of the popular Team Delta boards and some of the currently popular speed controllers have their own software failsafes built in. If you are using an Innovation First speed controller with a PPM radio, your failsafe should be adequate for the safety inspection of a competition.

IPD Radio Controller

One other branch of FM radio controller deserves a mention. *IPD* is similar to the PCM type. It is claimed that IPD is better than PCM because there is no *lockout*. When a PCM radio receives several false packets in a row, it engages the failsafe mechanism. When this happens, your robot will not do anything if the failsafes are set correctly. This is called lockout. IPD codes the packets of information but does not lock out when a lot of bad packets are received. Instead, it stretches the last good pulses. In short bursts of bad packets, you might do well to have an IPD-equipped radio controller. I know builders who have used the IPD, PPM, and PCM radio controllers: some have had excellent results, and some have had horrible results. The best thing for the new guy to do is buy the radio that allows expansion, is at least in the FM band, is within the budget, and obeys the rules of the competition. Then test it extensively.

Channels and Control Frequencies

Remote systems have what are called *channels*. Figuratively, a channel is the communication line that carries the command from you to the receiver. Each joystick on your transmitter can be moved around in all directions. Broken down, there are only two parts to the movement of the joystick. Vertical movement is associated with one channel. If you push the stick away from you in a vertical motion, the channel is telling whatever device on the receiver to go forward at a certain speed or to a certain position. Horizontal movement is associated with another channel. If you push the stick side to side in a horizontal motion, the other channel is telling whatever device on the receiver to go forward at a certain speed or to a certain position. During normal operation, the two channels never mingle and you can run two separate motors at two separate speeds by using the channels.

Some channels also refer to the *frequency* on which your transmitter (TX) and receiver (RX) are operating. Don't confuse these two types of channels. The different frequency channels available in the United States are listed in Appendix C in the "American RC Channel Frequencies" chart. Most competitions require that you be able to change operating frequencies in case there is another robot on the same one. Depending on your radio system, you change frequencies in one of two ways. All systems contain *crystals*, which dictate the frequency of your system. The crystal is installed in a socket in the receiver so that it may be changed out. To change the frequency of the receiver, grasp the plastic tab and give it a firm pull. It should pull right out. Put your new crystal back in its place. Now the transmitter must be changed. Some transmitters have the same type of setup as the receiver. If yours does, simply repeat the process of changing out crystals. Other transmitters have a module in the back that must be pulled out and changed. Still other transmitters have small dials on the side that can be set to the correct channel (frequency) so that it matches the frequency of the receiver. Any time you work on your remote control transmitter or receiver, you should take the bot out for a test drive. Make sure you are getting about the same reception as you did before. Do not wait until you are in the arena to test out the new crystals.

Some bots have many functioning parts that require many controls. It is possible to need more control channels than a standard radio can provide. In this case, you can buy a nonstandard RC setup that contains many channels and control the bot using that. Vantec and others sell modified transmitter/receiver pairs that increase the number of channels available. You can also use two transmitters and two receivers to control your bot, but you will have to make sure that both TXs and RXs are on different control frequencies. You must also notify the competition that you require two control frequencies, so that they can ensure there are no conflicts. If two or more functions can be controlled in the same exact manner at the same exact time, you can use one channel to control them by connecting the inputs to the same channel output of the receiver. You must make sure your receiver can supply the current necessary to run the control electronics for the extra functions. (Normally, it will.) It is also possible to have a bot in which one half of it is completely, electrically isolated from the other half. In this case, you need at least two receivers, one for each side of the bot. It is not necessary to have two transmitters to control two receivers. One transmitter can control both receivers if all three are on the same control frequency.

Tank Steering

In many two-wheeled robots, one drive motor is connected through the speed controller and whatever other interface if any, is necessary, to channel one on the receiver. The second drive motor is connected to channel two, which is normally controlled by moving the right joystick in a horizontal motion. This gives the illusion that you are controlling both motors together. Actually you are controlling both motors separately, since you have to concentrate on the vertical and horizontal positions of the joystick. To move a robot configured in this manner straight forward, you must move the joystick forward and to the right, the same amount at the same exact time. Otherwise, one motor will move faster than the other and cause the robot to veer to the right or left. You can make it easier on yourself by moving the control inputs for the drive motors on the receiver so that the horizontal or vertical channel of the right joystick controls the right drive motor and the horizontal or vertical channel of the left joystick controls the left drive motor, as shown in Figure 2.2. This method, in my experience, is slightly easier to deal with while driving.

You can make it much easier to control your robot by adding an *elevon mixer*. This device takes its input from channel one and channel two of the receiver and "mixes" those inputs in a special way to control the robot in much the same way as a remote control car. The physical mixer in many computerized RC units has been replaced by software. In that case, consult your RC unit's manual on how to turn on and adjust the elevon mixing.

Using the mixer is really easy. Most of the time it is simply a matter of installing the device and then doing some driving practice. A few relatively inexpensive units are sold by hobby shops, and circuits are also available on the Internet for the do-it-yourselfers. I've always had success with the Ohmark brand. An elevon mixer makes your two channels seem

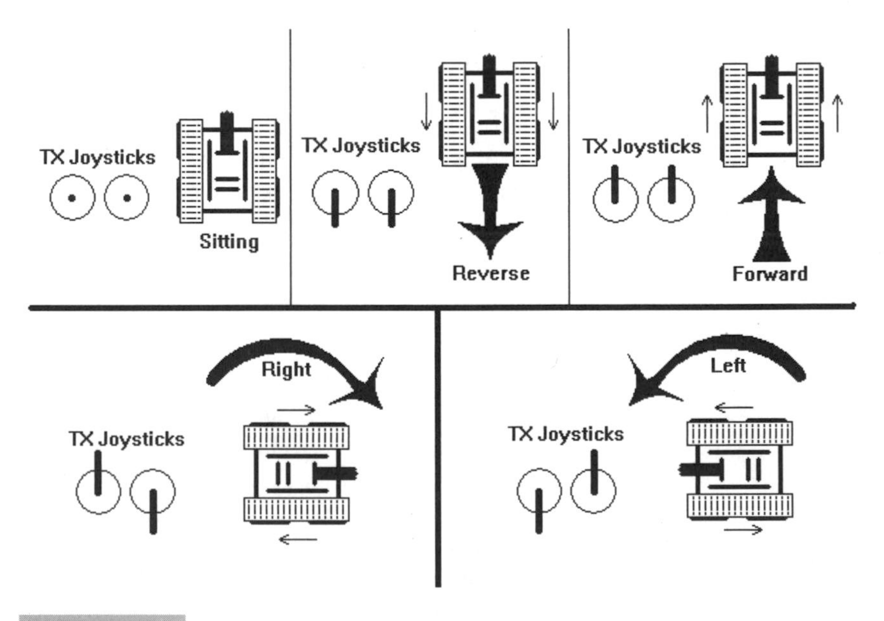

FIGURE 2.2 **Non-mixed tank-style steering with two joysticks.**

like one, which controls both motors at the same time. When you push the stick forward, vertical only, both motors run forward at the same speed as commanded by the position of the stick. When the stick is pulled all the way back, both motors run in reverse at top speed. When the stick is all the way to the left, the right motor spins forward and the left motor spins in reverse. This makes the robot spin to the left on its wheel axis. Moving the stick to the right produces the opposite effect and enables the robot to spin completely around in a zero turning radius. Most mixers also allow for arcing turns. If the stick is full forward and slightly to the left, both motors are turning forward but the left motor spins at a slower rate than the right. The bot then arcs toward the left. This same control also works in reverse.

A *gyro* is a useful object to some bot builders. The first couple of robots I built were two-wheel drive with a swivel caster that supported some weight. Because the caster could swivel, even the slightest power differences between the drive wheels would be greatly amplified in the bot's direction of travel. A gyro, as used in model helicopters, reduces the amount of turning that a two-wheeled robot is likely to display and helps make the robot go in a straight line. It does this by reading the signals being sent to the speed controllers, reading the direction and rate of spin of the bot, and altering the signals so that, if the bot is veering to the right, power increases on the right side to try and straighten it out.

RC Problems

The biggest problem with controlling your robot will probably be remote signal reception or getting the commands to the robot. The three main causes of problems are RC battery power, a badly installed antenna, and radio frequency interference.

BATTERY POWER

The first problem is overcome simply by making sure you have both your transmitter and receiver batteries charged before going into battle. To simplify it even more, the receiver batteries can be eliminated altogether. To do this, have the receiver draw its power from some other source in the robot. Most receivers normally run on a four-cell *NiCd battery* pack that comes with the RC setup. NiCd cells have a 1.2-volt rating. Four NiCd cells wired together as they come from the factory combine to give 4.8 volts total. Many electronic devices are designed to use a range of positive supply of voltages. Your receiver should be able to handle the extra two-tenths of a volt if you power it from a 5-volt supply. This is fortunate for two reasons. First, NiCd batteries, when fully charged, hold more than 1.2 volts. Thus, the receiver must naturally be able to handle more than 4.8 volts. Second, there are many ways to get 5 volts from just about any robot power system.

One specific way, and probably the easiest, is to buy a *DC-to-DC converter*. The wiring diagram is shown in Figure 2.3. This device simply inputs a voltage higher than 5 volts and outputs exactly 5 volts. You can build one with a voltage regulator, but I find it's easier and worth the relatively small amount of extra money just to buy one. Another advantage of using a DC-to-DC converter is that it normally has *ground isolation*. This means that the ground or negative side of your power source to the receiver is separate from the battery or whatever you are using to get the power. This prevents nasty ground loops that can toast your equipment and hampers the noise created by motors and other actuators on the receiv-

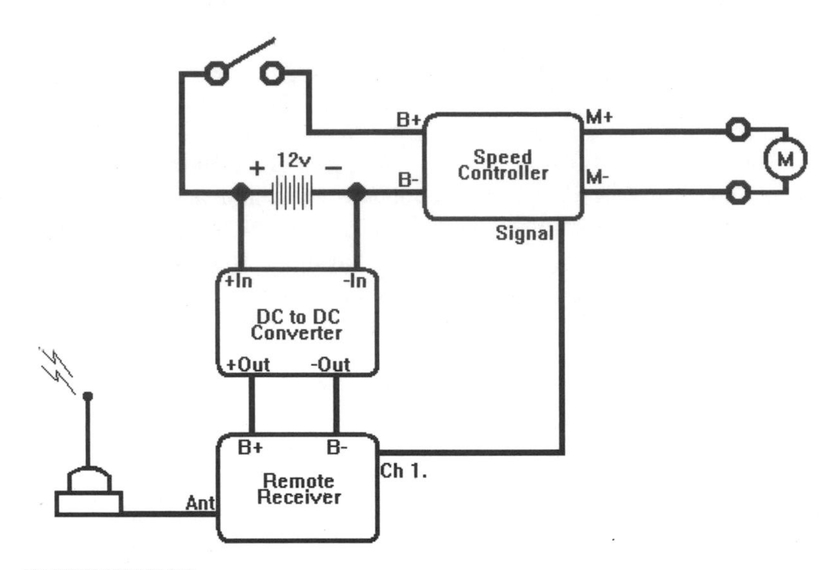

FIGURE 2.3 **Using a DC-to-DC converter to eliminate small receiver batteries.**

er's power lines. When buying a DC-to-DC converter, make sure you get one that can do two things. First, it must operate on the supplied voltages in your robot: many can operate on a range of voltages for the input, and some can even supply a range of output voltages. Second, it must be able to supply the amount of current demanded by all the devices that it powers. This is not only the remote control receiver but also any interfaces that draw power from the receiver. The manufacturer of those devices will tell you how much current they demand. For the most part, all the devices you work with on your first few bots will be OK when powered by a converter that puts out 300 milliamps (mA) or more. Team Delta sells some nice battery eliminator boards that contain everything needed to convert the supply voltage to something your receiver can use.

ANTENNA STUFF

Many people experience failure in the arena because their antennas aren't mounted properly on the robots. I once didn't qualify for a competition simply because my antenna wire had been accidentally cut and I lost control of the bot. Antenna mounting is more than the physical mounting of the device. It also includes the length of the antenna. Radio signals travel in waves. The length of the wave dictates how long the antenna should be. Ideally, your antenna should be the same length as one wave of the signal it's trying to pick up. The *wavelength* for a 72-MHz system is 4.167 meters (about 164 inches). The wavelength for a 75-MHz system is a little shorter, at about 157 inches, but a 150-inch antenna isn't really feasible for a combat robot. Fortunately, radio waves can also be plucked from the air nearly as efficiently at quarter length waves. One quarter of 4.167 meters is 1.042 meters (about 41 inches). This is a much more manageable antenna length. You can also use a half wavelength if you like, but it's still a bit uncontrollable at 82 inches. The closer to full wavelength the antenna is, the better it works.

Mounting the wire antenna is somewhat of a black art. I've seen many different ways of routing the wire. Some people run it taped on the inside of metal or plastic armor. Some run it taped on the outside of metal or plastic armor or rolled around a hollow or solid piece of wood or plastic. Some run it through a flexible plastic tube, as on an RC car. The best thing to do is experiment with your own system, remembering that for best reception it should be mounted vertically. Mount the antenna several ways, and see which one works best for you. Keep in mind that the antenna is the lifeline between you and your bot. If it gets cut or damaged, you may lose control. If you have followed the rules, your bot should stop once the radio connection is dead. This is called being disabled and is how your opponent claims victory.

Replace the wire antenna. Some people, including me, like to use an after-market RC antenna called a Dean's base loaded whip. Most of these consist of a hard piece of wire, about 6 inches long, connected to a small circuit (base load) that electronically makes up for not having the full quarter wavelength antenna (see Figure 2.4).

This type of antenna all but replaces the wire that dangles from the receiver. Some directions to follow when connecting it are:

1. Cut the existing antenna wire and leave between 4 and 10 inches on the receiver.
2. Assemble the antenna connector.
3. Mount the antenna base to the robot.

You must make sure you don't mount the base within an inch of any metal, graphite, wiring, or radio equipment. This means you can't drill a 1-inch hole in the metal top of your robot. It must be 2 inches in diameter for the antenna to be 1 inch away from any

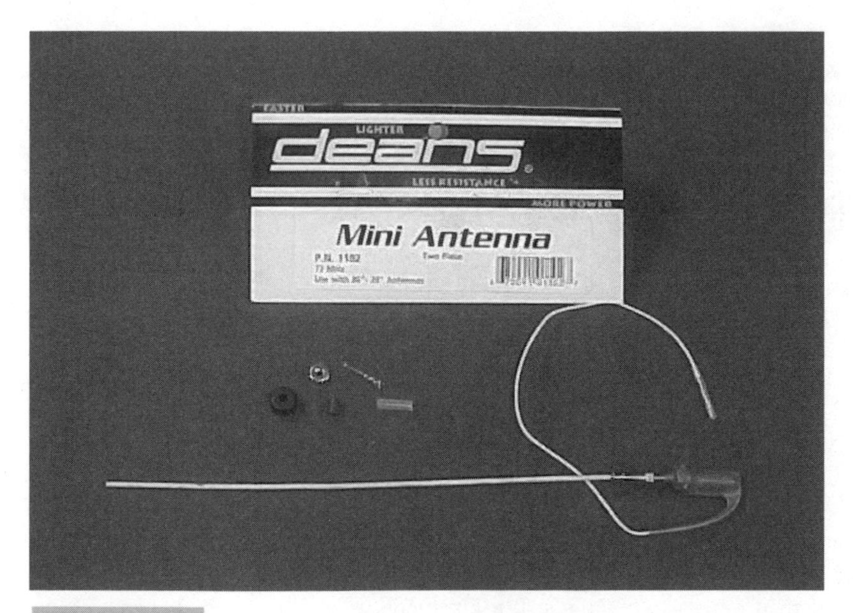

FIGURE 2.4 **Dean's base loaded whip antenna.**

metal. However, I mounted one of my whips in a 1-inch hole without problems. It really depends on how far away you will be from the robot while it is operating. Also, these antennas are not extremely sturdy, because they are made for model helicopters and other noncombat vehicles. It is a good idea to protect the whip antenna with some nonmetallic armor, such as a sizable piece of polycarbonate.

Figure 2.5 shows one of my whip antennas. It is mounted inside two pieces of 1/2-inch thick polycarbonate. First, I cut the polycarbonate the appropriate size. Next, I used a router with a small cutting bit to make a slot down the center. Then I used a larger bit to make a larger slot to accommodate the base of the antenna. This particular antenna was used on three different robots competing in five competitions. In the picture you can see some of the damage the polycarbonate sustained during a fight, but the antenna was not harmed. This mounting system keeps the antenna in a horizontal position, which degrades the reception quality. However, because drivers are usually relatively close to their machines, the signals get through pretty well.

Replace the transmitter antenna. I have one suggestion about an antenna for the transmitter. I use a rubber-coated, flexible antenna called the Rubber Ducky antenna instead of the metal, telescoping antenna from the manufacturer (see Figure 2.6).

I like it for a couple reasons. It has a quick disconnect at its base. Also, I've seen plenty of the telescoping antennas bent or broken. The telescopes can also put out an eye if you aren't careful. Some competitions are held in an enclosed arena that has a metal frame. Touching your telescoping antenna to that frame can introduce lots of radio interference. Most of the time you won't even know you did it until it's too late.

The rubber antenna is slightly more complicated to install than the base loaded whip of the receiver. First, you are required to open up the transmitter. Usually six to eight screws on the back of the transmitter are removed to get it open. Once open (see Figure 2.7),

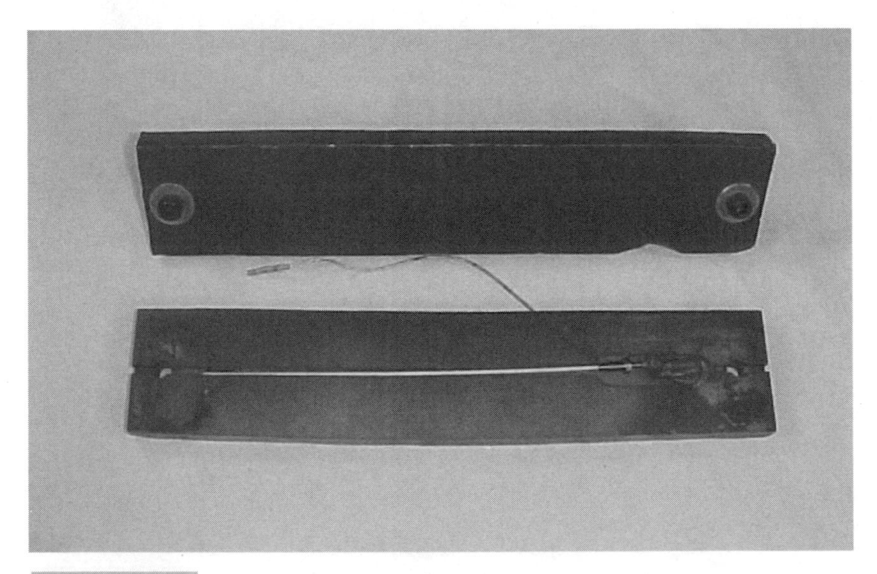

FIGURE 2.5 Mounting a base-loaded whip antenna inside polycarbonate.

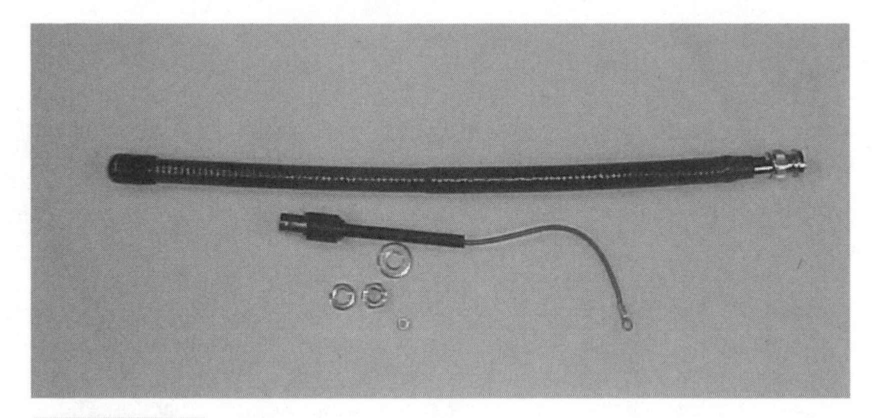

FIGURE 2.6 Rubber Ducky antenna.

unscrew the factory antenna. Some have a threaded stud that the antenna screws down on; others have a tab with a threaded hole that the antenna screws down into. The rubber antenna base connector has a hollow threaded rod attached. The wire to the quick disconnect goes through the hollow rod and is attached to where the telescoping antenna was mounted. The Rubber Ducky antenna should have mounting hardware and directions for either type of connection. Once you have the antenna installed, take your robot for a spin and recheck the reception.

FIGURE 2.7 Rubber Ducky antenna mount.

MOTOR NOISE

You are going to battle more than your opponent in this sport. *Radio frequency interference* (RFI) is your second enemy in the arena. Most people use permanent magnet motors in their bots and, because of the nature of these beasts, they create sparks that create RFI. You might think it's a good thing to create something naturally that could screw up your opponent; actually, it's bad because it can mess you up as well and because it's against most competition rules.

Capacitors. The first line of defense against RFI is a *capacitor* (part N in Figure 1.1). A capacitor is similar to a small battery in that it stores an electrical charge. However, do not try to run your bot off one. Capacitors either block or allow electricity to flow through depending on the frequency of the electricity and the size of the capacitor. The smaller the capacitor, the higher the frequency it will swallow or make disappear. It takes a good bit of expensive and specialized equipment to detect the frequency of the noise generated by motors, because usually more than one or two frequencies are generated. For these reasons we have to guess at what size capacitor to use and test its effectiveness in killing the radio noise problems. Start with a capacitor between 0.1 μF and 0.01 μF. When you buy capacitors for this purpose, pay attention to the voltage ratings. Your bot's system may be 24 volts, but voltage spikes that are much higher can be produced. Try to find capacitors that have a voltage rating of at least three times the motor voltage or more. If you don't, you will end up replacing the capacitors fairly often. Capacitors can also come with *polarized* leads. Because you will be switching the flow of electricity to switch the spin direction of the motor, you do not want these capacitors. It is easy to tell the difference: these *electrolytic capacitors* mark the negative lead with a minus (−) sign. Connecting an electrolytic capacitor to a power source in the opposite direction can cause the capacitor to explode.

Now that you know you need a capacitor, you need to know what to do with it. There are a couple different things to try depending on whom you ask. One group says to use three capacitors per motor. This is done by placing a capacitor between the brushes and one from each brush to the motor housing. The other group says to use only one capacitor, between the brushes, per motor. Both ways are shown in Figure 2.8. In fact, it's claimed that you can introduce a high current directly to the frame of your robot if you use three capacitors. I have always used three capacitors per motor and have never witnessed the "high current on the frame" situation, although their supporting argument seems logical. I leave it to you to try each configuration and determine which helps you the most. I know a few builders who have never used capacitors on their motors and never had problems. If you go that route, remember the capacitor when your bot starts freaking out because of interference.

Figure 2.9 should give you a good idea about the mechanics of mounting capacitors to a motor. I cannot cover every type of motor, but the main thing to remember is that the capacitors should be as close to the brushes as possible. If for some reason you cannot open the motor casing, go ahead and install the capacitors on the outside. Figure 2.9 shows two examples of motor capacitor mountings. The first example shows the motor case opened to install the capacitors. This is a fairly large motor and the caps can be mounted directly on the brush housing. The second example shows where a single cap was mounted on the outside of the motor.

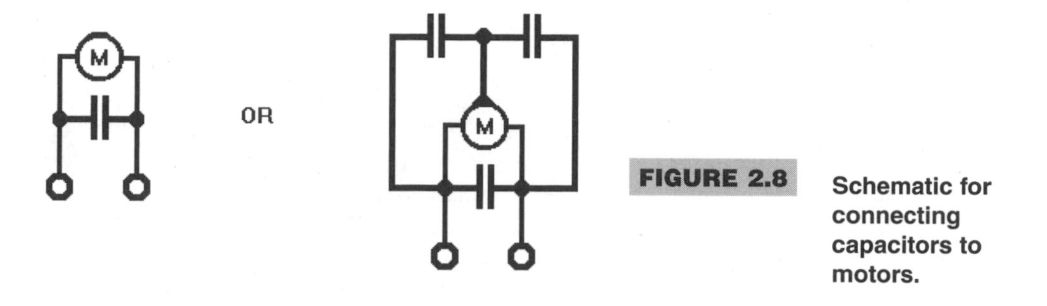

FIGURE 2.8 Schematic for connecting capacitors to motors.

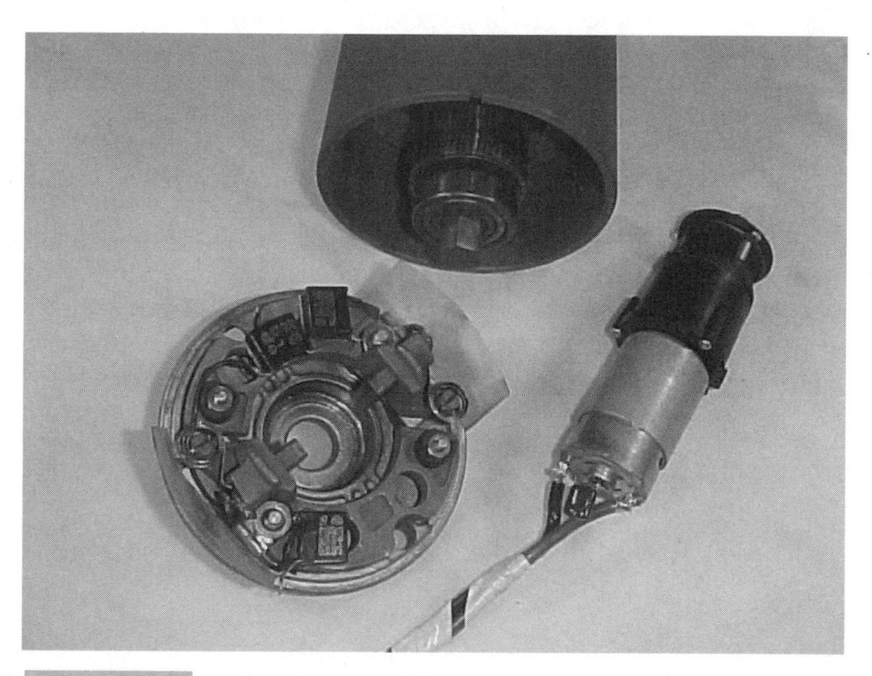

FIGURE 2.9 Installed motor capacitors.

RFI killers. Several other things can be done to reduce RFI. You can twist all the power wire pairs. Each motor has a pair of wires leading out. Twisting the wires creates a capacitorlike situation and can help with RFI. Ferrite cores can be used, but it is very difficult to determine exactly where to put them and which size to use. You can route wires through braided steel to help cut down on the amount of RFI that radiates out of the wires. One of the best—and sometimes hardest—things to do is to keep control circuitry away from power-delivering wires. However, you can place the circuitry inside its own metal housing. Aluminum housings are easy to use and lighter, but offer less protection than steel housings.

FCC Regulations

The last thing I have to say about RC setups concerns the law. Two very popular bands of hobby remote controls use 72 MHz and 75 MHz. The 72 MHz band is reserved by federal law for remote controlled aircraft. The 75 MHz band is reserved for everything else on land or water. Using a remote control system in a manner inconsistent with the laws is punishable by a $10,000 fine. The problem is that the 75 MHz band RC setups don't usually have as many control channels available as the aircraft band RCs. So, for several years, most people who built battling robots used the aircraft band and ignored the law. I have not heard of a case in which the FCC has pressed charges against a builder or an event promoter for using the aircraft band, but now that the sport is gaining immense popularity, anything is possible. It is becoming more popular to find a 75 MHz system with enough channels or to buy a 72 MHz system and have it converted to the land frequency. Conversion is a fairly quick and inexpensive procedure if your entire setup can be done, but some receivers cannot be converted. This leaves you two choices if you want to comply with the law: you can have your transmitter converted and buy a new 75mHz receiver, or you can sell your entire setup and buy one that works on the lawful frequency. It's up to you to decide, but I strongly suggest compliance with the laws.

If you decide to convert your system, be sure to ask how it's being tuned and to which section of the frequency band. You must have this information to buy extra crystals and get optimum performance. Each radio is tuned on a particular frequency number. If the person who converts your radio tunes it for the upper end of the channel spectrum, you want to buy extra crystals in the same end of the channel spectrum. A model airplane enthusiast told me not to buy crystals more than five channels away from the tuned channel. Because of the distances between the airplane and the driver, model airplane pilots need the best performance they can get. Usually, however, robot drivers are fairly close to their machines. I'm not sure how much of a difference it would make to use channels at either end of the spectrum, but I still tend to follow the guy's advice.

Summary

In this chapter, we explored the differences between radio systems and the importance of these differences. We also found that steering a robot like a tank is the easiest and most common method for driving a robot. This applies even if the bot has eight wheels. We discovered that the remote control system is the lifeline between you and your robot. It must be free from problems for your robot to obey your commands.

In Chapter 3, we talk a bit about the different types of wheels, along with the advantages and disadvantages of each type. We also delve into the question of how many wheels your robot should have. I'll even touch on walking robots.

3

REACH FOR THE FLOOR

Wheels are a big part of your bot. A strong drivetrain is absolutely necessary to a strong robot. After all, immobility is the measure of defeat in these contests. Choose your wheels carefully: you depend on them to move you around and push others around. Figure 3.1 shows examples of *pneumatic* and solid wheels. The large black one is a pneumatic go-cart racing tire. The two gray ones are the popular Colson caster, solid wheels. The smaller black one is a generic utility cart wheel.

Pneumatic Wheels

Pneumatic wheels are a good choice for many bots. They are usually strong and very grippy on the battle surface. You also have an adjustment that can be made to get more traction or more speed—you can let the air out or put more in. Letting air out increases the *contact patch* of the wheel and decreases the radius, whereas pumping it up decreases the contact patch and makes the wheel taller. Increasing the contact patch of the tire may or may not help traction. Often there is a lot of debris on the battle surface that tends to degrade traction. However, letting air out of the tire also decreases its *radius*. (In Chapter 6, you will see that using smaller tires increases the amount of torque available. You will also see how putting more air into the tire increases the top speed of the robot by increasing the diameter of the tire. These are slight adjustments and in no way should you plan on getting large

FIGURE 3.1 **Different types of wheels.**

benefits from doing either. You should always buy, design, and build the drivetrain so that it does what it's supposed to do.)

Pneumatic wheels also usually come with a hub that can be easily mounted to the driven axle of your bot by way of a bolt pattern. The air-filled wheel has some disadvantages, too. They can be punctured or torn easily in battle. Some competitions have spikes, saws, glass, or fire coming up from or lying on the floor. Foam-filled pneumatic wheels do very well against these types of hazards, but they can be heavy.

Solid Wheels

Colson caster wheels are another veteran bot builder favorite. Caster wheels are usually made of solid rubber, plastic, or other material. The best kinds for bot building are those that have a hard, durable hub and a strong but grippy rubber bonded to the hub. I'm not talking about the little black ones at the local hardware store. These wheels are fairly light and come in a lot of different sizes. Their disadvantage is that the hubs are sometimes difficult to mount to the driven axle because of bore size. They also suffer from traction losses a lot sooner than pneumatic tires.

Cheap lawnmower wheels from the local hardware store are right out. These are usually made of plastic with some hard rubber tread. They are difficult to mount to an axle because of the thin hub material, and they lack traction and durability. It's funny that many new builders cut costs by using lawnmower wheels even though the drivetrain is the most important part of the bot. I have used these wheels and lived to regret it, though the bot did not.

Number of Wheels

Two-wheel drive bots are less complicated to build but also more difficult to drive. You may wonder how a robot with only two wheels steers. The answer is simple: each wheel has a motor. You control the motors independently through the remote control. When both are going forward, your bot is going forward. When one is going forward and the other going in reverse, you are turning your robot away from the forward moving wheel. This is called *tank steering* because this is how a tank moves.

Four-wheel drive bots are more difficult to build because you've got four motors or you have to have some way to transfer power from one motor to two wheels. However, because there is more robot on the ground, a four-wheeled bot is much easier to control compared to the two-wheeled bot. To steer a four-wheeled robot, some robots employ automobile or riding lawnmower–type steering, where the two front wheels turn at the same angle. Most bots don't use this method. Most bots have a mechanical or electrical connection between the two wheels on the right side and a connection between the two wheels on the left side. This keeps the front and back wheels on each side moving in the same direction at the same speed. These bots use the tank steering method, just like a two-wheeled bot. This method is sometimes called *skid steering* because to turn the wheels skid or slide across the floor. This usually gives excellent straight-away control over the robot while increasing the turning radius.

Six-wheel drive bots are even more complicated to build than the four wheelers, but you also get the advantages of getting more power to the ground in different situations. Some six-wheelers have a hard time steering because of the increased skidding of the tires. To overcome this, you have a couple choices. Putting more powerful motors in the bot may seem like the best choice but you do have a weight limit—you simply may not have room within the limit to add stronger motors. The second choice is to lower the center pair of wheels by a small amount. This gives the robot a slight teeter-totter effect, lifting either the front or the rear pair of wheels off the ground slightly and making it easier for the wheels to slide across the floor.

Again, the type of wheel is more important than the number. You need a wheel with a strong hub and a soft(ish) outer contact ring. Hard plastic wheels do not get the traction that you need. The bigger the better is not always the rule either. I used 5-inch wide go-cart racing slicks and still did not have the traction I wanted. I used 3-inch wide, pneumatic, knobby cart tires with the best success. Depending on which type of competition you are entering and your bot's design, you may not want to get your tires filled with foam. It is an easy decision: if the competition has arena hazards that can puncture or cut your air-filled tires, they will go flat and you will lose; if your bot design allows your tires to be vulnerable to attack, they will go flat, and you will lose.

Some bots use track mechanisms instead of wheels. Tracked bots, like everything else, have advantages and disadvantages. One advantage is that it's hard to stop a tracked vehicle if it has the proper traction. A tracked bot might ride right up on top of a wheeled bot in a fight. On the bad side, tracked bots are fairly complicated to build and operate successfully. You have only a couple choices when it comes to tracks. You either have to buy them or build them. Buying them can be expensive. Even if you buy, you have to adapt them to the rest of your robot. When looking for tracks, try checking businesses that sell snow blowers and other outdoor power equipment. Some people use sprockets and an attachment chain to build tracks; others prefer to mill the tracks and supporting wheels from raw aluminum or steel. If you build tracks, spend the time and money to do it right. The added complexity doesn't lend itself to reliability, so if you do a shoddy job, it will fail.

Walking Bots

Walking robots are some of the most spectacular bots you'll see. They are also the most complex and difficult to build. Building legs that can pick up and move the weight of your robot while getting bashed by your opponent is a considerable task. Based on the fact that the Honda corporation has spent billions to get a two-legged machine to reliably balance and walk, I venture to say that no one is going to build a walking, fighting robot with two humanoid legs. Translated, that means I don't think anyone will build a version of C3PO and put him in the ring with combat robots.

TWO-LEGGED BOTS

Even so, there have been a couple two-legged or *bipedal* fighting robots. These bots have two feet: one foot is a big, hollow ring or square. The other foot is a big plate that sets in the middle of the ring. The ring-foot raises off the floor and moves forward. The plate-foot stays within the ring and on the floor while the ring-foot moves. The ring-foot lowers back to the floor, and this time raises the weight of the robot, including the plate-foot, off the ground and moves it forward. Combining many of these two motions gives the robots forward and reverse walking control. Turning is accomplished by spinning the weight of the bot while the ring-foot rises.

WALKING GAIT

The pattern and order in which the feet of a walking robot touch the ground are called the *gait pattern*. The gait pattern of the two-legged example above is very simple. One foot is always on the floor and the feet alternate. Such a bot could balance on one foot, because the feet are big enough to support the center of gravity of the robot. There have been four-, six-, and eight-legged fighting robots as well. The gait pattern for these bots can be very complex or very simple. The challenge is to find actuators and a gait pattern that move the legs fast enough to be totally effective in battle situations.

Until recently, walking robots were at a disadvantage when pitted against wheeled robots. In many competitions, walkers were given additional weight limits. This weight limit extension was meant to aid the walking robot builder because it is more difficult and takes more material to build sturdy, true walking robots. Even with this weight allowance, walking robots were at a disadvantage when fighting a rolling robot—that is, until *shufflebots* came along.

SHUFFLEBOTS

Some people call the shufflebot a walker and some people do not. In my opinion, the shufflebot doesn't meet the spirit of the requirements to be considered a walker. A shufflebot rotates a long foot by means of a cam or some other device. When several of these long feet are placed side by side on two sides of the robot and rotated rapidly out of phase, they propel the robot in a shuffling manner. Because of the weight allowance that was meant for robots with actual legs, these shufflebots were able to increase the power of their weapons and the strength of their armor without requiring the weight for real legs. When this happened, the major competitions changed the rules so that shufflebots wouldn't have an automatic advantage over the rest of the field.

Summary

In this chapter, you found that the choice of wheels can make the difference between winning and losing. However, there is no clear winner among types of wheels—many bots use pneumatic wheels successfully, and many bots use solid wheels successfully. Each type has its pros and cons that must be weighed in the final design of your robot.

In Chapter 1, I said that electricity is the lifeblood of your robot. If that is true, speed controllers must be the beating heart of your bot. Next up, in Chapter 4, we find out what the most popular types of speed controllers are and how they work. We discuss their shortcomings, advantages, and some specifications. We even talk about how to take care of these important and expensive components of your robot.

4

SPEED CONTROLLERS

Speed control is a major aspect of any combat robot. The three basic methods for controlling the speed of an electric motor are the "Bang Bang" method, *variable current,* and *pulse width modulation* (PWM).

Bang Bang and PWM

The "Bang Bang" method of speed control really isn't much speed control. The motor is either completely on or completely off. This method is the cheapest and easiest way to implement control of a robot. However, it makes the bot very difficult to drive because you can switch the power on and off only so quickly. The variable current method requires some way to—you guessed it—vary the current the motor sees. This requires some type of *variable resistor* and a way to remotely control the setting of the resistor. This method gives a lot more control over the speed of the motor yet has its own drawback: because current is escaping the motor through the resistor, the design wastes a lot of electrical power. That power can be put to much better use inside the motor itself.

PWM is a technique of changing the voltage the motor sees by varying the width of a pulse of voltage to the motor. It sounds complicated but it really isn't. Basically, the pulse is either on or off. While the pulse is on, the motor is trying to run at full power. While the pulse is off, the motor is slowing down and eventually stopping. If you turn the pulse on and off fast enough, the motor's electrical characteristics fool it into thinking it is receiving a lower voltage. That makes the motor run slower. The drawback to PWM is that the controllers that do this technique are pretty expensive. Also, they are more fragile than other types of controllers. When using a PWM controller, be sure to protect it from vibration and impacts.

Real Speed Control

Several brands of PWM speed controllers on the market are currently being used successfully in combat robotics. Each has advantages and disadvantages. Figure 4.1 shows a Vantec RDFR22 on the left, a 4QD Pro 120 on the right, and a 4QD 300-amp controller with *tachometer* board in the center.

4QD CONTROLLERS

4QD, a British company, makes several controllers using between 12 and 48 volts that are capable of controlling a range of currents between 0 and 300+ amps. These controllers need a special, relatively inexpensive, interface to decipher the remote control receiver's signals. A few interfaces are commercially available and a few web sites tell how to build them yourself. These controllers typically control only one motor, so you'll have to have at

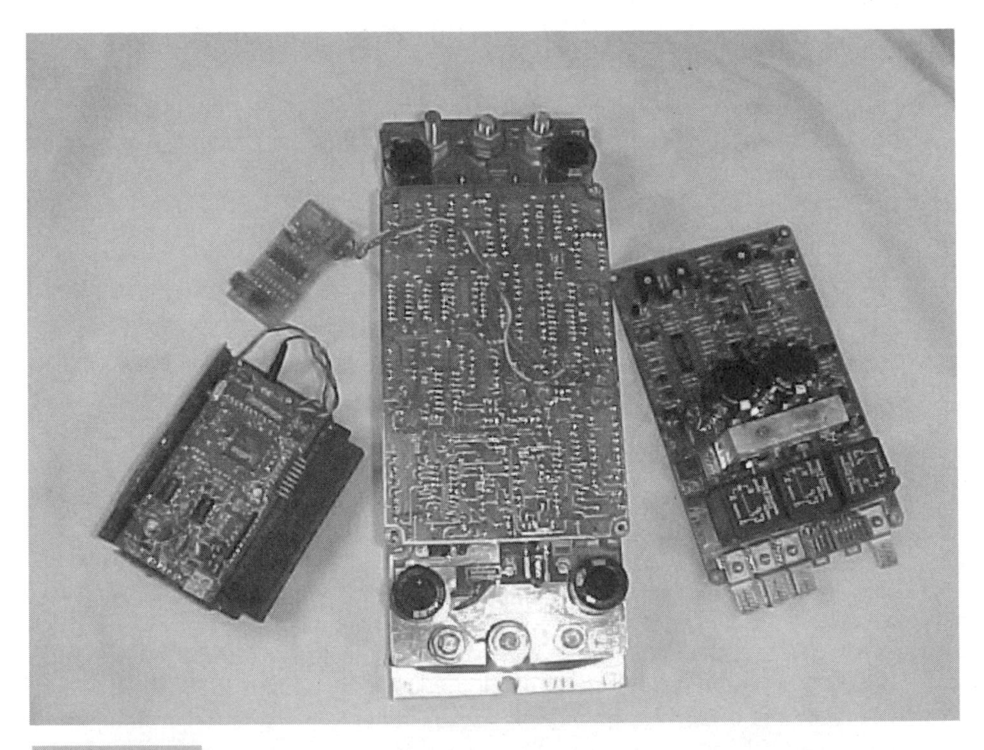

FIGURE 4.1 4QD and Vantec speed controllers.

least two. 4QD controllers have numerous features including *current limiting*, acceleration and deceleration ramping, and high pedal lockout, just to name a few. Current limiting is nice because it protects the electronics in the controller from trying to supply too much current to a big motor. If that happens, the electronics in your expensive speed controller are toast.

Acceleration and deceleration ramping is a feature that protects your motors, speed controllers, and other drivetrain components. While in combat, you are constantly changing directions. Some changes of direction induce violent changes of motor spin direction. When this happens, large voltage spikes can fry electronics, and large torque forces can destroy gears, chains, bearings, and any number of drivetrain components. Ramping doesn't let this happen. When you give the command to change direction, the 4QD controller slows the motor down before changing directions. The ramping feature can be adjusted so that it takes a couple of seconds to slow down and change directions or so that it takes only a few hundredths of a second.

High pedal lockout is more of a operational safety feature. It prevents the robot from tearing across the floor at top speed when you turn the power switch on and have inadvertently pushed the transmitter joystick trim into running position.

VANTEC

Vantec, a California-based company, also makes several controllers that use voltages ranging from about 5 to 60 volts and are capable of controlling currents ranging up to about 220 amps. These controllers have a built-in interface that decodes the signal from the receiver. They also have a function known as *signal mixing* or *elevon mixing*. Most PCM radios have this mixing function built in, so it's not necessary for the speed controller to have it, but sometimes it is nice. Vantec controllers can also usually control two separate motors at once. These features make the Vantec controller one of the most used controllers in the combat robotics community. However, because these are very popular, make sure you order your controller well in advance of the tournament.

INNOVATION FIRST

A strong, new competitor in the speed controller market is Innovation First, Inc. A Texas-based company, they build compact and reliable controllers specifically for robot control. They currently only have 12- and 24-volt versions and can handle 90 to 110 amps comfortably, although their spec sheet says 60 amps continuous. They don't have mixing capability either, but they are some of the more robust and easy to use controllers on the market. They are also currently the least expensive controllers. Look for them to start adding more features as they get their feet wet in this new market.

OSMC

A very recent competitor to the market is the open source motor controller (OSMC) unit. The OSMC unit was designed by several motor controller experts within the combat robotics community. Its features include mixing, flipped bot motor reversal input, upgradeable and customizable software, serial interface, and network capability. All the schematics are available online for free, as is all the software. Or, you can purchase entire units online through Robot-Power.Com. It's sort of like buying a copy of Red Hat Linux—Linux is free but you're buying a version that has been tested and upgraded from the standard open source. This is the essential nature of open source material.

RC CARS

Last, you have remote control race car speed controllers. Most of these cannot handle the current levels that motors demand in a large robot. Only the beefiest controllers will be able to handle the smallest robots. Most of these controllers do not control the direction of the motors either. They simply go forward. These controllers are the cheapest controllers on the market but not by much. It's much better to save your money and buy one that will do everything you need it to do now and in the future.

Care and Feeding of a Speed Controller

You can do a few things to try to protect your equipment. First, a visual inspection of the board itself is a good idea. Look for loose parts or solder bridges that can cause failures ranging from flaky operation to down right up-in-smoke failures. Vantec, 4QD, and OSMC have open case designs and are easy to inspect. Innovation First controllers are enclosed in a plastic shell and sealed with warranty tape. Breaking the tape breaks the warranty, so a visual inspection is up to you.

Because Vantec, 4QD, and OSMC units come with an open case or no case at all, you should fabricate something to enclose the controller so that no stray metal can get to the circuit board. Use the cover while building, testing, and competing.

The circuit boards are also susceptible to vibration. You should mount the controller so that outside vibration or shock from fighting collisions is minimized. Rubber mouse pads make good vibration insulators.

Read the manual and don't exceed the voltage or current ratings. Doing so is a sure way to let the magic smoke out. The easiest way to protect against high current is to install a fuse in series with the batteries and the speed controllers. Suppose you have a speed controller that can handle 60 amps continuously and 120 amps for 10-second pulses. If you know that your bot should not draw more than 70 amps during its hardest load, you can install an 80-amp fuse. The fuse should blow before the speed controller exceeds its 120-amp rating. This becomes less and less true when you start using fuses close to the controller rating. Controllers usually have MOSFETs or similar electronic devices to conduct the current to the motors. These devices burn much more quickly than a fuse can pop, so it's probably not going to do you any good to put a 120-amp fuse in line with the 120-amp peak controller.

Fuses are a matter of personal taste during a fight. Most people prefer to use them only while testing the robot. The thought behind this is that nothing matters except winning the fight: if you are going to fry your speed controls or wiring or motors, you are probably going to lose the fight anyway. Why take the chance of losing because of a weak fuse? There is always the possibility that your controllers will experience only a very short term of overrating and still recover, giving you the extra 5 or 10 seconds needed to stay in or even win the match.

Summary

In this chapter, we talked about speed controllers. I personally like the 4QD controller for all of its features. However, the OSMC unit may have the most features for the best price. If you want to keep everything simple, and they can handle the current requirements, I believe the IFI controllers will get you rolling more quickly than anything else.

Chapter 5 covers batteries. I'll show you the most powerful and the most popular types of batteries used in combat robotics. We'll talk about charging them, and I'll show you how to wire them up to get the voltage and robot runtime you desire.

5

BATTERIES

Lots of people don't know the specifics of the types of batteries that are normally allowed in fighting robot competitions. Because of the chances of the robot turning upside down or of the batteries being punctured, batteries that have a liquid *electrolyte* (standard car, motorcycle, and lawnmower batteries) are not usually allowed and are never a good idea. When battery acid spills, it makes a mess of a lot of things including everything on the inside of your bot. The acid can cause serious burns to people, too.

Sealed Lead Acid (SLA)

Currently, one of the most common types of battery used in combat robots is the *sealed lead acid* (SLA) or *gel cell*. There are slight chemical and manufacturing differences between these types, but we consider them the same for our purposes—with the exception of one particular brand. Hawker or Odyssey brand batteries are top of the line and have several characteristics that make them the best choice among SLAs. More on that comes later in this chapter. SLA batteries have electrolyte that does not spill when the battery is broken, cut, or punctured. Some have a high current output, too, making them ideal for the short run, high power drain situations that occur in a fighting robot.

SLAs are also probably the cheapest way of powering your bot. I've paid $5 to $35 for 12-volt surplus batteries on the Internet. Surplus batteries usually aren't the best batteries you can buy in terms of discharge current capability but they can do the job at the expense of weight and power. You can sometimes find surplus batteries at flea markets and Hamfests that are cheaper than on the Internet or in stores. I've also heard of one builder getting them free from a hospital, when maintenance had to change all the batteries in the emergency lights and had no use for the perfectly good used ones. Good deals can be found, but be sure the battery can hold a charge and can put out the amount of current you need.

In Figure 5.1, you see the duration of *discharge versus discharge current graph* for a Panasonic LC-RD1217AP, 12-volt, 17-*amp-hour* (Ah) battery. By looking at the graph, you can see that you should be able to get about 17 amps for up to 10 minutes before the battery voltage drops below 12 volts. If you notice the curves for 34 and 51 amps, you will not get a full 12 volts from this battery when drawing those currents. You can overcome this problem by wiring two batteries in a parallel circuit. (I will explain parallel batteries in the Multiple Batteries section later in this chapter.)

The Hawker name brand is hard to find sometimes, but Hawker also makes the Odyssey brand battery with the same specifications. These are targeted to the motorcycle and Jet Ski market. They have a much higher discharge current capability than a normal SLA. Along with the higher discharge current capability comes a higher price. I've paid between \$65 and \$150 for new Hawker/Odyssey batteries. Prices vary from dealer to dealer on the Internet or in a store. Figure 5.2 shows the performance data of an Odyssey PC545 13Ah SLA battery.

The Odyssey can put out about 1200 amps when it's short-circuited, although you really don't want to do that. The chart lists information about how much current can be discharged over a specified amount of time. SLAs are rated like any other battery, by using the amp-hour. If you have a 13-Ah battery, you might expect to be able to draw 13 amps for 1 hour continuously. This isn't actually the case: many batteries, like the Panasonic, are rated over a 20-hour period. In the case of a Hawker, the batteries are rated over a 10-hour period. You will actually get about 1.3 continuous amps over a 10-hour period from a 13-Ah battery. In other words, you can draw 1.3 amps from a PC545 for about 10 hours before it is putting out only 10 volts. Ten volts is considered dead. Once you try to drain a normal SLA battery faster than its rated time period, you get only about 40 or 50 percent effi-

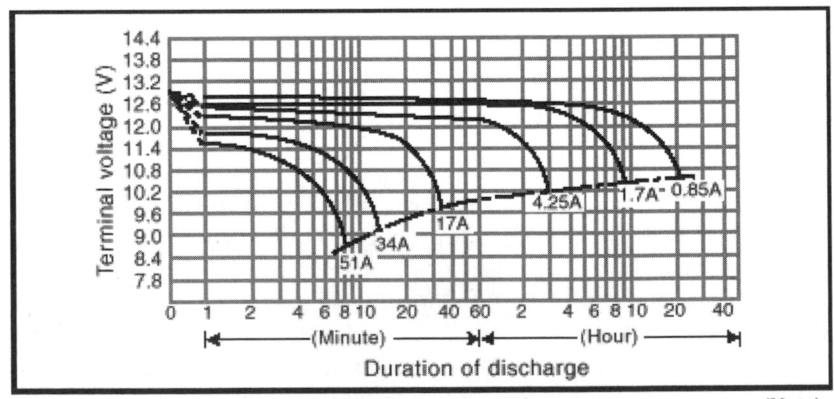

FIGURE 5.1 **Panasonic battery discharge graph.**

ODYSSEY® PC 545 performance data at 25 °C, per 12V module

Time to 10.02V	Watts (W)	Amps (A)	Capacity (Ah)	Energy (Wh)	Energy and power densities			
					W/lit.	Wh/lit.	W/kg.	Wh/kg.
2 min	1268	123.9	4.10	42.30	665.20	22.20	264.10	8.80
5 min	758	70.8	5.90	63.20	397.90	33.20	158.00	13.20
10 min	482	43.6	7.30	80.30	252.80	42.10	100.40	16.70
15 min	361	32.2	8.05	90.30	189.50	47.40	75.25	18.80
20 min	292	25.7	8.60	97.20	153.00	51.00	60.75	20.25
30 min	214	18.6	9.30	106.80	112.10	56.00	44.50	22.25
45 min	154	13.2	9.90	115.65	80.90	60.70	32.10	24.10
1 hr	121	10.4	10.40	121.20	63.60	63.60	25.25	25.25
2 hr	67	5.7	11.40	134.40	35.30	70.50	14.00	28.00
3 hr	47	3.9	11.70	140.40	24.60	73.70	9.75	29.25
4 hr	36	3.0	12.00	144.00	18.90	75.55	7.50	30.00
5 hr	29	2.5	12.50	147.00	15.40	77.10	6.10	30.60
8 hr	19	1.6	12.80	153.60	10.10	80.60	4.00	32.00
10 hr	16	1.3	13.00	156.00	8.20	81.85	3.25	32.50
20 hr	8	0.7	14.00	168.00	4.40	88.15	1.75	35.00

FIGURE 5.2 Odyssey performance data for PC545 battery.

ciency. Hawker batteries are a little better than that, but I always use the same efficiency numbers as a safety blanket of available battery power. So, you actually only get about 5.2 continuous amps for 1 hour before the PC545 is dead. Your robots aren't going to be fighting for an hour at a time; most competition matches have time limits of 5 minutes or under. To find out how much current the PC545 will put out for 5 minutes, multiply the 1-hour ability, 5.2 amps, by 60 minutes. You get 312 amps for 1 minute. Divide that by 5, and you get 62.4 amps over 5 minutes. So, if your robot will not draw more than 62.4 amps continuously and runs on 12 volts, a single PC545 should power it.

MULTIPLE BATTERIES

What happens if you design your bot to run on 24, 36, 48 or any other voltage? Now we get to add more batteries into the equation. I'm sure you've noticed that the numbers above are multiples of 12—$2 \times 12 = 24$, and so on. If you want 24 volts, you wire two batteries together in *series*. Thirty-six volts requires three batteries and so on. Wiring two batteries in series means that you connect one of the battery's positive terminals to the other's negative terminal, as shown in Figure 5.3. Then you use the two loose ends as the battery connection, treating the two batteries as one. Measuring the voltage across the loose ends gives you about 24 volts. To get 36 volts, put another battery in the line. The reason I say you get "about 24 volts" is that, when fully charged, batteries carry a higher voltage than their 12-volt rating. You may actually see 26 volts or more from two fully charged batteries. For simplicity's sake, we will just say that two batteries add up to 24 volts and so on.

Now suppose that your robot needs 120 amps to run continuously for 5 minutes and uses only 12 volts. One battery isn't going to provide the needed current to last the entire

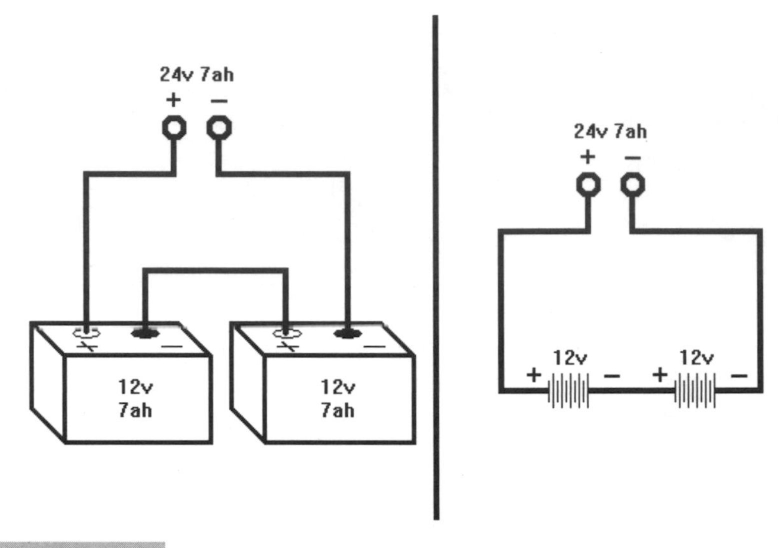

FIGURE 5.3 **Series battery connection diagram.**

match. In fact, it will last only about half the match. You have two alternatives: beat your opponent very quickly or add a battery in *parallel* with the original. Parallel wiring means that you connect both batteries' positive terminals together and then connect both batteries' negative terminals together, as shown in Figure 5.4. Measuring the voltage across the loose ends gives you 12 volts, but the current lasts twice as long or you can get twice as much in the same period of time. To make it last three times as long, you add another battery to the parallel lineup.

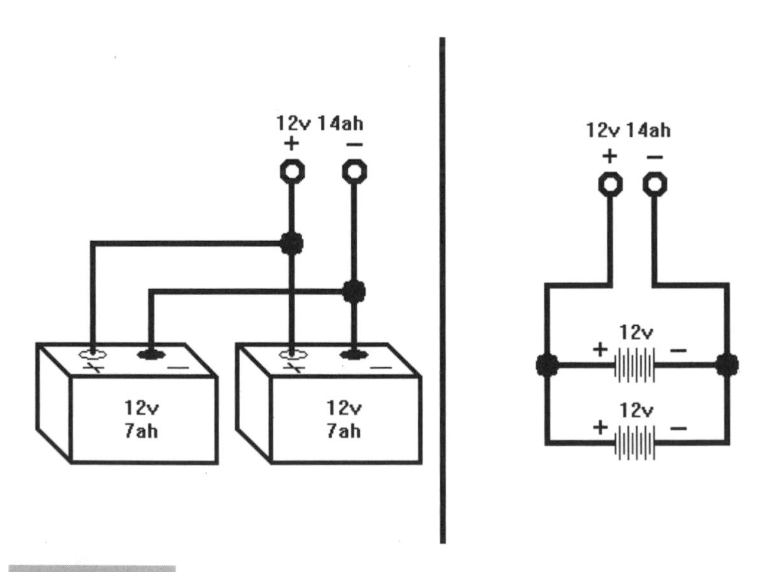

FIGURE 5.4 **Parallel battery connection diagram.**

Another possibility exists. Suppose your robot needs 120 amps to run continuously for 5 minutes and uses 24 volts. One pair of batteries wired in parallel gives you the current necessary but will not supply the 24 volts. One pair of batteries wired in series gives you the voltage necessary but will not supply the amperage needed. The solution is a combination of series and parallel wiring. Think of the parallel-wired batteries as one 12-volt battery with twice the current capacity. Think of the series-wired 12-volt batteries as one 24-volt battery with the same current capacity as a single, solitary 12-volt battery. Now, if you wire two 24-volt series packs together in parallel, you get a 24-volt battery that can handle the required 120 continuous amps for 5 minutes. It may be clearer in Figure 5.5.

Using series and parallel wiring, you can form any multiple of 12 volts at many different, continuous amp ratings as long as the weight limit and your cash allow. SLAs also come in different voltages. Because of different discharge ratings and different charge ratings, you should not mix batteries of different voltages, or you can end up destroying an expensive piece of your robot. Once you have more than one 12-volt battery in your robot, you have access to all the different voltage levels as well. If you have three 12-volt batteries wired in series to get 36 volts, you can tap one of them and get 12 volts. You can *tap* two of them and get 24 volts. Figure 5.6 shows two batteries wired in series to get 24 volts while one is being tapped for 12 volts. I recommend this only if what you are powering on the lower voltage doesn't draw a lot of current. The tapped batteries run out quicker than the untapped, which can cause problems and possibly destroy the batteries or at least drop their efficiency even lower than normal.

Now that we know about multiple batteries, lets look at the Panasonic and Hawker batteries again. If we wired two Panasonic batteries in parallel, the same curves in Figure 5.1 apply to the two batteries. The difference is that each battery will supply half the current that you draw and last twice as long as a single battery doing the same job. So, if you wire two

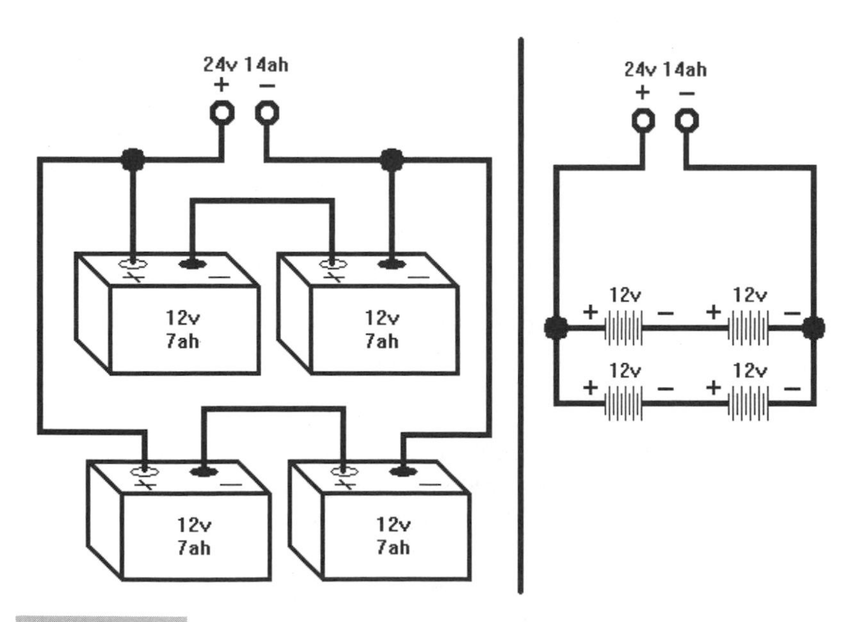

FIGURE 5.5 Series-parallel battery connections.

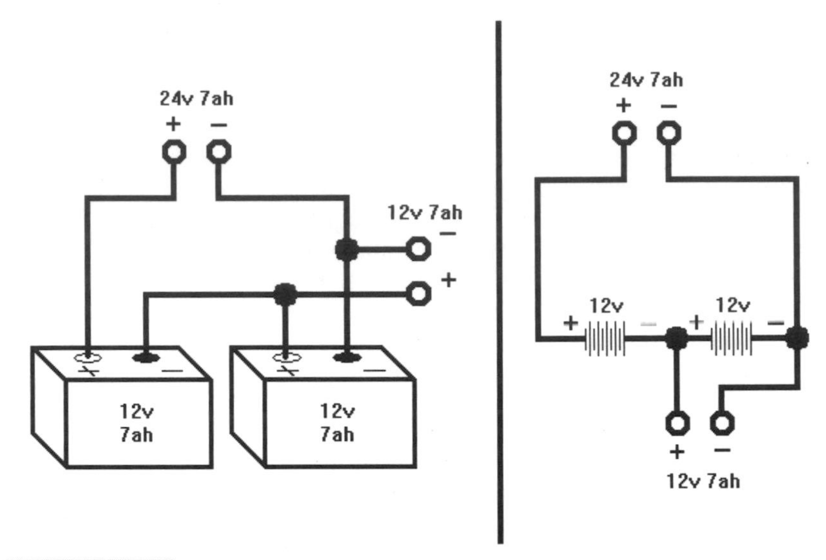

FIGURE 5.6 Getting two voltages from series wired batteries.

of these batteries together in parallel, you should be able to draw about 34 amps at 12 volts for about 10 minutes. If you have three batteries in parallel, you should be able to draw about 51 amps at 12 volts for about 10 minutes. Some speed controllers can handle some voltage sag without shutting down but the limit is usually around 10.5 volts, which is why some people can use one of these batteries to supply 51 amps at 10.5 to 11.5 volts—that and the fact that the battery can supply the 51 amps at 11.5 to 12.5 volts for almost a minute. The voltage drops really fast, though. Remember that voltage and current dictate how much power your motor is putting out. Also remember that each of these batteries weighs about 14.34 pounds. Three in parallel weigh about 43 pounds. You can get the same 51 amps at 12 volts for about 8 to 9 minutes out of a single Odyssey PC545 13-Ah battery for about 12.5 pounds. Panasonic and other brands of SLAs have their place, but if you need big current, you need the Hawker/Odyssey brand or lots of NiCd or NiMh batteries.

CHARGING SLAs

Charging common surplus or new SLAs is a little different than charging the Hawker/Odyssey SLAs because of internal differences. With either type, proper and adequate charging is the single most important factor in obtaining the optimum life of the battery. *Constant current* (CC), *constant voltage* (CV), and a combination of both are three standard methods for charging these batteries. CC chargers are more expensive and complicated and therefore not usually used for charging SLAs in our type of application even though they generally provide shorter charging times. The combination method has the same downfalls as the CC chargers.

CV charging, where a single voltage level is applied across the battery terminals, is the most suitable method to charge an SLA battery. It also happens to be the manufacturer-recommended method of charging. The CV charger is capable of supplying a constant voltage and a varying amount of current depending on the charge state of the battery. In other words, the battery determines the amount of current it draws from the charger, up to the limit of the charger.

The standard SLA has a limit to the charge current that it can safely handle. It's usually almost half (0.4CA) of the amp-hour rating on the battery. Also, the voltage that you charge the battery must be in the range of about 14.5 to 14.9 volts for a 12-volt battery. Many people use standard car battery chargers to charge standard SLAs and the Hawker/Odyssey type. Be careful to pay attention to the voltage and number of amps going into the battery. Many automobile chargers put out only about 13.8 volts. Your battery will charge, but not completely. These chargers are designed to bring an empty car battery to a state of charge that is high enough to turn an engine. Once this is accomplished, it is up to the engine alternator to bring the battery to a full charge by topping the battery off with a trickle charge. Find the battery's data sheet from the manufacturer to see the best way to finish off the charge. You can find suitable power supplies for charging these batteries on the Internet. However, most dealers will be able to point you to a charger that will get the job done.

The Hawker/Odyssey type SLAs have an advantage over the standard SLA, because they have no in-rush current limit. These batteries can handle as many amps as your charger can supply as long as the voltage of the charger stays within the manufacturer-prescribed range of 14.7 to 15 volts for a 12-volt battery. I have charged mine with a 200-amp charger, but I felt more comfortable charging at the 40-amp setting. To get a good charge on these batteries, your charger should bring the battery up to 14.7 volts and then switch to a *trickle charge* of 13.6 volts.

Regardless of which type of SLA you have and of which type of charging scheme you use, you must put between 1.05 and 1.10 amp-hours into the charge for every amp-hour discharged to ensure a complete and adequate recharge.

Just a quick note on something I've found to be helpful when trying to charge batteries at a competition. I have more than one charger. Each of my chargers only charges a single 12-volt battery or 12-volt pack of parallel-wired batteries. If I have a 24-volt pack made of two 12-volt series-wired batteries and don't have a 24-volt charger, I charge them both at the same time with two 12-volt chargers. You run the risk of not charging one or more batteries to the same extent, but you are sometimes pressed for time. Or you may choose to have one or more extra sets of batteries so that you can remove the used set between matches. You replace them with the fresh set and put the used set on the chargers for as long as necessary. Ideally, you should have at least three sets of batteries per robot. One in the bot, one fully charged and waiting, and one on the charger. After all, why design a bot with much more battery than you need to make it through a match? Do the math. Figure out how many batteries you need and use the extra weight for a better weapon or more armor.

NiCd and NiMh

The next most common battery type is a *nickel-cadmium* (NiCd) battery. (I include the *nickel-metal hydride* [NiMh] batteries along with NiCds for any further discussion purposes.) Both types are common in the RC racing car industry. They have high current output capability, and are also lighter and more flexibly mounted than SLAs because they come in small cells roughly the same size as regular alkaline batteries. Some builders buy a lot of single cells and build their own custom voltage, custom amp-hour, custom shape packs that fit perfectly into their robot. Some builders buy the 6- or 7-cell packs that are used in RC cars. There are even companies on the Internet that build custom NiCd packs for bot builders. One of the most notable brands of packs is Hilltop Batteries' BattlePacks.

Two drawbacks must be dealt with when using NiCd batteries: they don't source as much current as an SLA and they cost more—usually between $2 and $8 per cell. (I've bought surplus cells for 50 cents each in quantities of 100.) These come in several ratings up to about 3 Ah or 3000 mAh (milliamp-hour). NiCds are usually labeled using the mAh specification. One major advantage the NiCds have over the SLAs is that you get about 90 percent (0.9CA) of the stamped amp-hour rating as opposed to 40 percent of an SLA when drawing lots of amps in a short time. NiMh cells have an 80 percent (0.8CA) rating. Sub-C cells are the most popular in the RC racing world, having had more research invested to produce a better, more powerful cell. For these reasons the sub-C is the most popular among the bot builders who use NiCds.

BUILDING NiCd PACKS

Unlike the standard 1.5-volt alkaline C cells, each NiCd cell carries 1.2 volts. They are wired in series to bring the voltage up to whatever is specified in the bot design. Then they are wired in parallel to bring the amp-hour rating up to specs. Manufacturers usually spot weld small, flexible tabs onto the ends of each cell to accomplish the wiring. Builders don't usually have the equipment to do this but that does not really matter—it is a bad idea to use tabs because they can break easily under the forces that are experienced during a match. Also, because the spot weld has a higher resistance than a solder joint, more power will be turned into heat instead of being used in the motors. Builders who make their own packs usually solder wire to the ends of the batteries to build a *pack*. Flexible, thick wire with lots of strands and a high-temperature insulation is the ideal connector between battery cells because it's harder for an impact shock to break the connection. Use a wire large enough to handle the needed current that the batteries can supply. This current is usually about 30 to 40 amps per pack, so you can use a 10 AWG wire between cells and be safe.

Solid mounting, even between battery cells, is a hazard for any electronics in your bot. In Figure 5.7, you see the beginnings of building a NiCd battery pack. To solder a wire to a cell, apply some *flux paste* and heat up the end of the cell for a very short period. Apply some solder as it is heating. Next, *tin* the wire by heating it up and applying solder. Clean your soldering iron tip and melt some solder onto it as well. Next, place the tinned wire on the end of the cell. Apply heat to the wire until the solder melts and forms a nice connection. A nice connection is shiny and smooth, not dull and chunky. Make sure the wire is in contact with the cell and not just lying on top of the first bit of solder. Be careful not to overheat the cell, because this will cause damage. Scuffing the ends of the cells with some fine-grade sandpaper can help make the solder stick. Here is where we see another drawback to NiCd batteries: if you solder tabs or braided wires to the ends to build the pack, the solder can melt under heavy current drain and cause a break in the connection. This problem can be overcome by making sure you have enough battery packs so that each pack provides only so much of the current load so that they do not overheat.

Once you have all the cells joined together to form the correct amp-hour and voltage ratings, you need to protect all those solder joints from shorting out on any metal objects within the robot. You also need to protect the pack from flying debris making contact

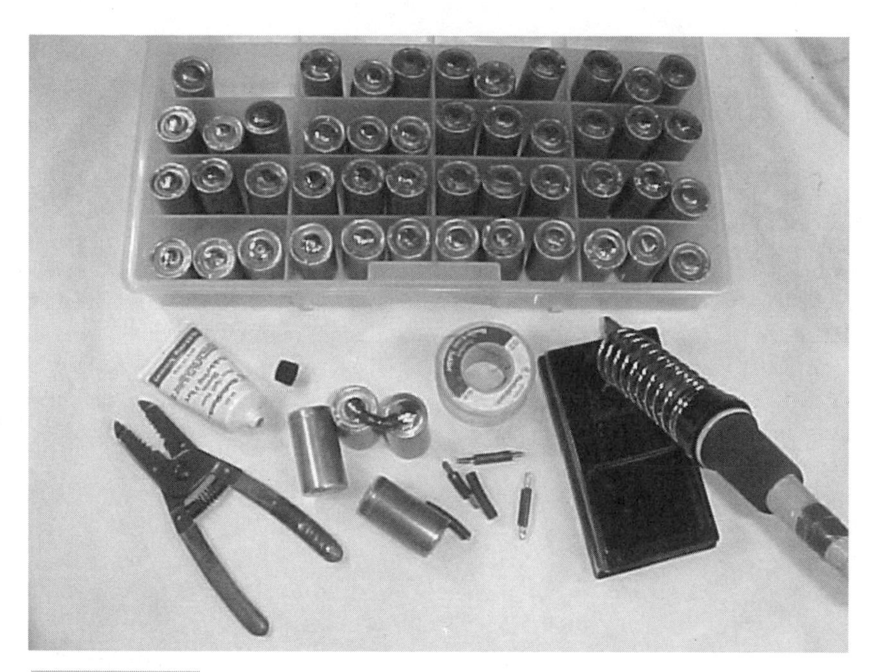

FIGURE 5.7 **Building a NiCd battery pack.**

between two cells that should have no contact. Most builders and businesses use *shrink wrapping* for this purpose. For those who don't know, shrink wrap is a type of flexible plastic tube that you put over the entire pack of cells. Once on, use a heat gun to heat up the tube. This causes it to shrink and form fit around the cells. All the RC racing car packs have shrink wrapping.

MEMORY EFFECT

There is some debate about the very existence of a NiCd memory effect. I personally have witnessed this effect, and only recently has it been made clear to me what is happening in the cell when this effect is apparent. First, it is not permanent. Using a cheap charger, it is easy to overcharge a NiCd battery pack. You can tell when this is happening by the temperature of the batteries. Although NiCds get warm while normally charging, hot batteries are being overcharged. The *electrode* in the cell experiences a—for lack of a better term—chemical hardening when overcharged. This causes the cell not to supply current for the entire specified time because the hardened area does not give up electrons as readily as the normal part of the cell. As time goes by, the hardened area of the electrode is used less and less because the rest of the electrode gives up its power more easily. You use the battery for a few minutes and put it back on the charger. Each time the battery is overcharged, more of the electrode is hardened. Each time the electrode is hardened, it appears that the cell holds less of a charge. This is called *shallow, repetitive cycling*. To reverse the hardening of the electrode, the cell must be completely drained and then recharged. This is called *conditioning*. Sometimes it takes several draining cycles to clear the entire effect. If you use, a good charger, you probably won't see the memory effect in a NiCd. If you do, simply make sure you completely drain the cells before putting the battery back on the charger, and try not to let them overcharge. The effect may not be enough to notice over a weekend competition, but in the off season take care of your batteries by charging them correctly.

CHARGING NICDS

Charging NiCds is a little trickier than charging an SLA. NiCds don't like to be overcharged. They get really hot and can even melt or explode. At the very least, you lose some performance from the cells. There are four common ways of charging NiCd cells: the voltage cutoff, temperature, timer, and peak methods. The voltage cutoff method checks the voltage of the pack and turns off the charger when a preset voltage is reached. The temperature method stops charging once a preset temperature is reached by the pack. The timer method simply lets the charger run for a specified amount of time.

Each of these has its own pros and cons, but none is as effective as the peak charger. When charging, a NiCd battery's voltage increases until the battery is full. Then it starts to overcharge, and the voltage of the pack starts to drop. A peak charger watches the voltage and stores the highest value in the memory. Once the voltage starts to drop, the charger is turned off. Many builders who use NiCds also use a peak charger from Astroflight. They seem a little expensive (at about $230), but the charger can charge up to 40 cells and the price includes a power supply that can source the necessary current. Although the price may seem high, many builders swear by them.

Flying Batteries

If you are building a fighting robot, you will most likely have to travel for a fight. At the current stage in fighting robot competitions, there aren't any regionals and having weekly competitions in your nearby city is probably not happening. If you are serious about the sport, you are going to have to travel. This may include flying with your robot or parts of it. Almost everything in a robot is safe to package in sturdy boxes and check as luggage, but there are a couple exceptions to this rule. Batteries and pneumatic tanks are on the airline's no-no list. Dry cell batteries are no problem as long as they are packaged so that they cannot cause a short. Wet cell batteries, like car batteries, are not allowed on planes, because they can spill too easily. Gel cell batteries are debatable. For a gel cell battery to get on a plane, there are a few guidelines that help. One, the battery and the outer packaging must be plainly and durably marked "Nonspillable" or "Nonspillable Battery." Two, the battery must be protected against short circuits and securely packaged. Three, the battery must be capable of withstanding vibration and changes in air pressure without leaking. The best thing to do is check with your airline for packaging specifications before buying tickets. Check with them several times. I called the same airline four times and gotten four different answers ranging from "You can carry them on" to "You can't bring that with you." Do not assume that because the battery manufacturers say the batteries are safe to take on a plane, the airline will allow it. In the worst case, you may have to ship your batteries using a separate shipping agent. Be sure to find out in plenty of time before the competition.

Summary

Batteries supply the life to your robot. Without them a robot is a boring, stand-still paperweight. The Hawker/Odyssey brand SLA batteries will give you the best performance when compared with other SLAs. NiCd and NiMh batteries are making their way into the robot combat community in a fast way. Correctly charging your batteries is vital to battery health and overall useful life.

In Chapter 6, we talk about the muscles—motors—of our robots. I offer ways to characterize different motors so that you can make comparisons and determine which is best for your robot. We'll even talk about how to gain strength through gearing.

6

ELECTRIC MOTORS

We're concerned mainly with weight, power, and ease of use when talking about motors. *Permanent magnet direct current* (PMDC) motors are the easiest to use for this application. The servos that came with your radio setup aren't strong enough for anything other than the 1-pound robots, like the one I build in Chapter 19, and even that weight class is steadily advancing past using servos. *Stepper* and *brushless* motors require a special driver. Series wound motors require special handling as well.

Alternating current (AC) motors require a device called an *inverter* to run off batteries. (An inverter converts DC electricity to AC electricity.) The voltage required for AC motors is usually higher than that for DC motors. The inverter is expensive and is, in the overall bot picture, wasted weight. All this and the added complexity make AC motors unsuitable for our purposes. With PMDC motors you can usually hook up two wires to get them running.

Choosing Motors

Everyone goes through pretty much the same steps when trying to find suitable motors to drive their bots. First, you look at your design and figure out what size wheels you want and what the speed of the robot should be. Next, figure out how much *torque* it will take to move the bot. Then, figure out what voltage your bot will be running based on the speed controller you will use. This calculation is a double-edged sword, because you also need a good idea of the current that the motors will draw, so that you can choose a speed controller. However, enough varieties of motor exist to allow you to go ahead and pick a speed controller that has been used by a lot of veteran builders and then choose the appropriate motors. Once you start experimenting with new exotic motors, you can find the speed controller that will handle them. As a beginner, you should stick to what's easy in order to get started. You must find a motor and speed reduction method combination that fits in the bot, fits within the electrical characteristics of the bot, fits within the weight restrictions, and gives you the speed and strength you require. To help you choose a motor that has been used by veteran bot builders, I have included the "Motor Spec Chart," Parts 1 and 2, in Appendix C, which gives specifications that you will need to know for many popular motors.

Gear Motors

The easiest PMDC motors to use already have gear reduction built in. Three popular geared motors come from wheelchairs, cordless drills (shown in Figure 6.1), and car windshield wipers. All of these operate on relatively low voltages and use amounts of current that are readily available from a battery. Each one has its different quirks when mounting them to the robot and mounting the wheels to the motor, but the main concern is whether or not the motor is capable of moving the robot.

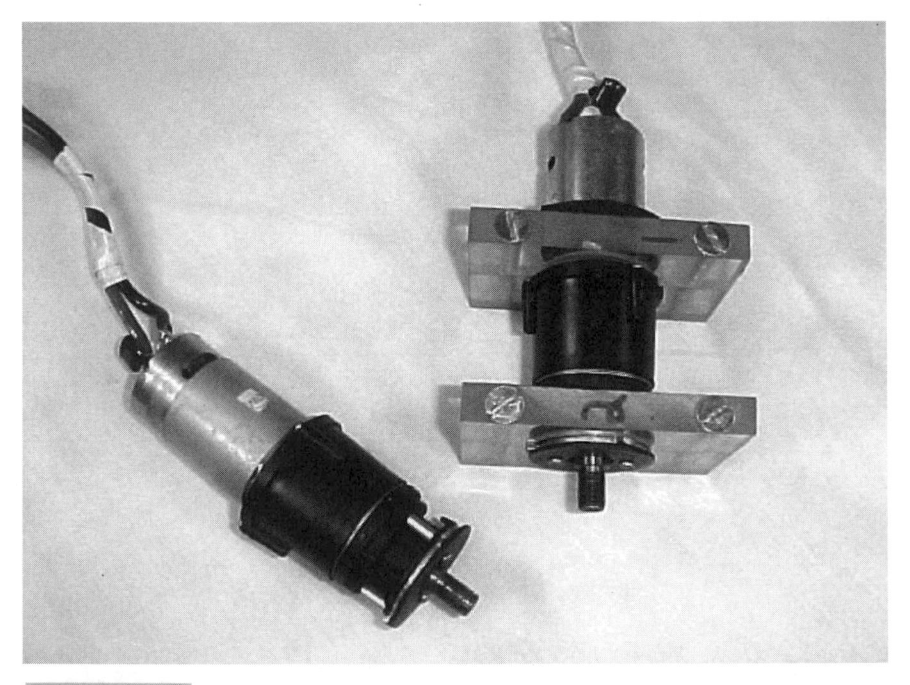

FIGURE 6.1 Gear-head motors from cordless drills.

Finding Torque

Most motors give the torque, voltage, current, and no-load revolution-per minute (rpm) ratings on the manufacturer's name plate. If one or all of these is missing, you can still determine the motor specs with a few simple measurements. The most important thing to figure out is the voltage rating: if there is no name plate, check out the manufacturer's Web site or give them a call. Try to figure out, based on looks, which motor it closely resembles before calling. If the motor came from old equipment, try to find the motor model from the manufacturer of the equipment. If you can't find the voltage rating, the only thing left to do is experiment. Start out with a small voltage. Try 6 volts. The motor may or may not spin. Continue with higher voltages until the motor starts heating up. Find a voltage where the motor seems to run comfortably.

Once you find the voltage rating, you can get the stall torque rating by running the motor at that voltage with the motor shaft locked with a wrench and pressing on a scale. When you do this, be sure the motor is securely mounted to the bench and the wrench is touching the scale. Do not apply the voltage for more than a couple of seconds. Prolonged stalling of the motor will destroy it. You then multiply the number on the scale by the number of inches from the point of contact on the scale to the center of the motor shaft. This will give you the *stall torque* of the motor in pound-inches (lb-in).

Look at the example shown in Figure 6.2, and suppose you have a 24-volt motor. On the shaft is a locking wrench. The length of the straight line between the center of the motor shaft to the point where the wrench touches the scale is 10.5 inches. The reading on

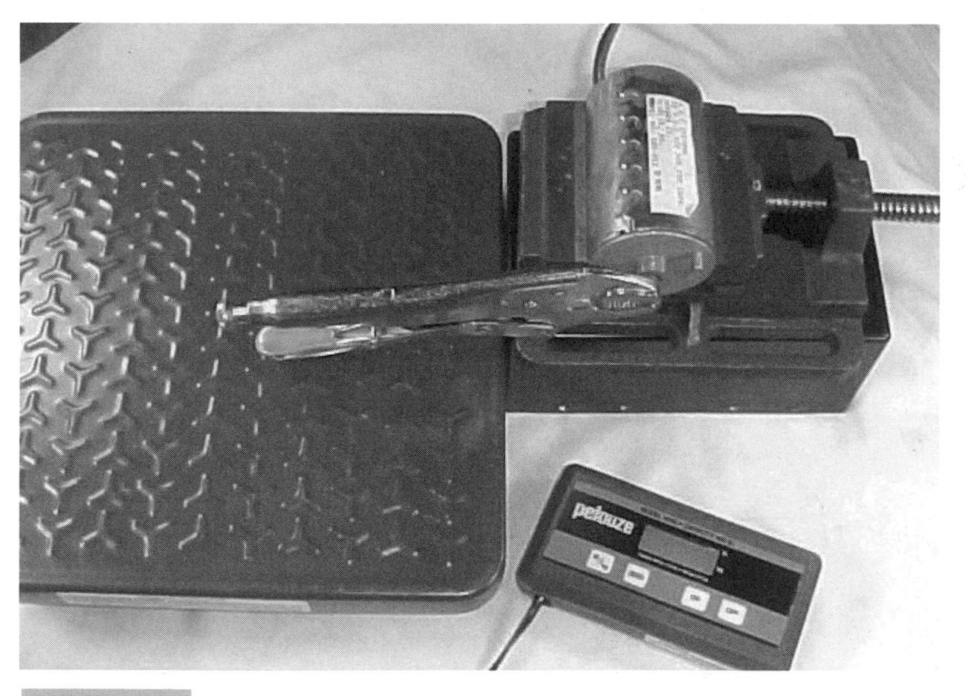

FIGURE 6.2 **Finding torque using a scale.**

the scale is 5 pounds, so 10.5 inches times 5 pounds is 52.5 lb-in. Hint for safety: test the motor's spin direction before applying the wrench. It should turn toward the scale. If not, reverse the motor power leads.

Motor Current Draw

To find the current that a motor draws, you can use a shunt, wired in series with the motor and battery, to measure no-load and stall current. A *shunt* is a calibrated resistor. You place the shunt in line or in series with the motor, as shown in Figure 6.3. Then you measure the voltage across the shunt and calculate the current being drawn through it using Ohm's law. Standard multimeters usually have the capability of measuring DC current, but most only measure up to about 20 amps. A 20-amp *peak current* measuring ability is usually not enough to determine the stall torque of a motor that is strong enough for use in a fighting robot. Another, although fairly expensive alternative is a *clamp-on amp-meter*. Some of these can measure thousands of amps accurately. Consult the owner's manual on how to use a clamp-on meter.

USING MOTOR RESISTANCE

You may want to skip the shunt and the clamp-on meter and use yet another way to find the current draw. Using Ohm's law, current (I) equals voltage (V) divided by resistance (R) or:

$$I = V / R$$

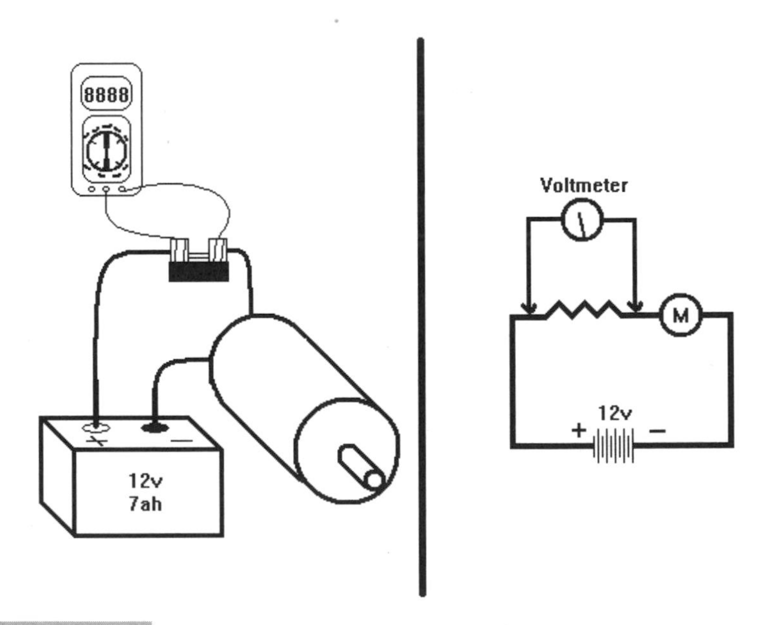

FIGURE 6.3 **Finding current draw by using a shunt and voltmeter.**

You need to find the resistance of the motor itself. You should not measure the resistance of the motor with a meter, because it is not sensitive enough. You need to find the resistance by substituting other values. Changing Ohm's law around, you get resistance (R) equals voltage (V) divided by current (I) or:

$$R = V / I$$

If we use a small voltage to run the motor, say 1.5 volts from a regular D-cell battery, and stall the motor, the current is low enough to measure with a regular amp-meter. The motor should not be allowed to turn (Figure 6.4). It probably won't, but the current on the line is correct. Now we have the V (1.5) and the I (measured current in amps) for the formula. Calculate the R. Now, put the R into the formula for finding current ($I = V / R$). Put the real running voltage in for V this time. Do not use the 1.5 volts from the regular battery. Now calculate the current (I) using those values. Your answer should be the stall current at your normal running voltage.

WIRE SIZE AGAIN

Stall current is useful in selecting motors, batteries, and speed controllers. The general rule of thumb is to design with stall current in mind. The only time I stray from this is when I'm choosing the size of wire for the robot. As I mentioned in Chapter 1, different wire sizes can carry different amounts of current. According to some sources, you need 4 AWG wiring if you want to carry 150 amps. That is true at certain distances for certain lengths of time, but the wire lengths and *duty cycles* we use in our robots are short enough so that we can use smaller wires to handle larger currents. The "Chart of Wire Size, Current, and Length" in Appendix C shows a range of wire sizes, lengths, and current-carrying capa-

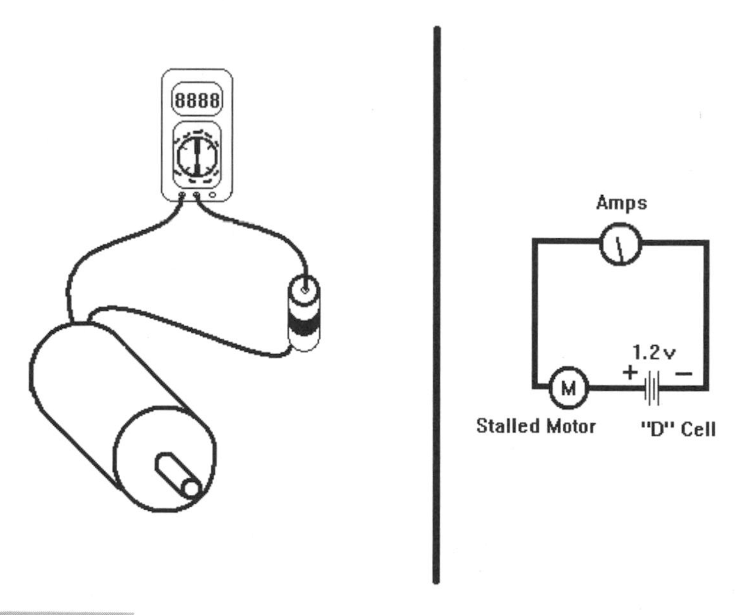

FIGURE 6.4 **Finding current using a D-cell battery.**

bilities. The "Chart of Wire Size and Current Capacity" in Appendix C shows the industry standard information. Quite a few differences in opinion arise between the two charts. Ultimately, it is up to you to decide which wire to use. I use the "Chart of Wire Size, Current and Length" to determine the size of wire I use, and I have never had a wire burn.

SPINNING WHEELS EQUALS LOW CURRENT DRAW

Along with keeping the wire lengths short, there is a reason we can run a motor whose stall current is 200 amps on a 10 AWG wire that should handle only 60 amps: we design our drivetrain to have more than two times the strength needed to carry our robot. Because of this and friction (or lack of it), the wheels start to spin when encountering forces something less than the total weight of the robot. In other words, the robot is strong enough not to make the motors stall. They spend most of their time in their comfortable running zone and drawing a good deal less than the stall current. If the motor ever does see a stall current, it's either for a very brief time or someone just destroyed part of your drivetrain.

Torque constants. A few things must be done to figure out how many amps your motors draw when your wheels begin to spin. Using the stall torque and stall current, you can find what is known as the *torque constant* (Kt) for a motor. The torque constant tells you how many amps it will take to produce a specified amount of torque. The formula for calculating the torque constant is:

$$Kt = \text{Stall torque (oz-in)} / \text{Stall current (amps)}$$

Kt is usually specified using ounce-inches versus amps, so be sure to specify the stall torque in ounce-inches. If your motor name plate indicates torque in pound-inches or pound-feet, use the conversion factors in Appendix B to convert them to ounce-inches. Some torque constants for popular motors are listed in Part 1 of the "Motor Spec Chart" in Appendix C.

Weight distribution. The weight of the bot is distributed across all the wheels and motors, but depending on the *center of gravity* of the bot, some wheels will have more weight on them than others. If you are still in the planning stages of building, you will not know exactly how the weight is distributed. Even if you have the bot in front of you, it can be complicated to figure out. For now, estimate the weight on the wheels by dividing the total weight of the bot by the total number of wheels touching the floor. This gives you the weight supported by one wheel. You could measure the weight supported by a single wheel by putting a scale under it, but the estimation is usually close enough. Convert that weight value into the torque that the motor must produce by multiplying it by the radius of the wheel and dividing the answer by the gear ratio. It may be clearer using the formula:

$$Sw = W / N$$

where:
Sw = Supported weight on one wheel
W = Total bot weight
N = Number of wheels on ground

$$T = (Sw \times R) / r$$

where:
T= Motor torque
R = Wheel radius
r = Gear ratio

Now convert the torque at the motor (T) into amps (A) drawn by the motor using the torque constant (Kt):

$$A = T / Kt$$

Depending on the actual weight distribution of your bot, you will be able to come close to calculating the theoretical maximum amperage your motors will draw up to the point when the wheels start spinning. If the amperage you calculate is higher than what your speed controllers can handle, you must change your gear ratio. After reading the Non-geared Motors section of this chapter and all of Chapter 7, "Gears, Sprockets, and Pulleys," check out the General Questions section of the FAQ in the back of the book for another example of calculating Kt and the amperage at wheel spin.

Motor rpm with a Tachometer

Finding the rpm of a motor is not difficult. The easiest way is to purchase some sort of *tachometer*. Many builders use a tachometer that is available from hobby stores and is intended to measure the rpm of model airplane propellers. These devices work by counting the times that a propeller blade blocks a light source pointing at the tachometer's sensor. You must rig some sort of propeller to the shaft of the motor you are working with. Standard propeller types have 2, 3, or 4 blades. Some tachometers can adjust their reading based on the number of blades. Other tachometers can't, and you will have to calculate the rpm based on the reading it gives. The user's manual for the tachometer model you own will show you how to do that.

A couple of safety and accuracy tips should be mentioned. Be sure to completely secure whatever you are using as a propeller to the motor shaft. Many motors run between 2,500 and 5,000 rpm or more. You do not want a piece of wood or plastic flying off the motor shaft toward you while measuring rpm. Also, secure the motor to the workbench in some way. Many of these tachometers are accurate up to 12 inches from the propeller. Stand clear of the spinning blade.

Tachometers measure how many times a separate light source flickers, and some can measure 32,000 rpm and higher. That's the same as a single propeller blade blocking light at about 533 times per second. House lights flicker at about 60 times per second so the tachometer can easily detect that and destroy the accuracy of the reading. Sunlight is the best source of light for the tach. However, a battery-powered flashlight is just as good. If you use a flashlight as a light source, you do not have to turn off all the house or shop lights to use the tach; just make sure your flashlight is pointed directly at your tach and the tach reads 0 rpm while the blade is not spinning.

Motor rpm with a String

Some people like to save a little money by doing a little extra work. I balance this tendency against the amount of time I will need to prepare for and do the work. If you want to save $30 and not buy a commercial tachometer, you can rig one up in the shop. You need a good length of string, a stopwatch, your motor securely mounted, and a battery to power it. You also want someone there to either operate the stopwatch or the motor. You need to know the length of the string and the diameter of the motor shaft. Do not use too much string because the diameter around which it is winding is significant in the calculation of the rpm. Too much string will wrap around the shaft and make it bigger for measurement purposes. Do not use too little string, or it will be difficult to start and stop the stopwatch and get an accurate time reading. Connect the string to the shaft of the motor so that it will wind up when the motor is turned on. Whoever is turning on the motor can lightly hold the string so that there is enough tension that it wraps neatly around the shaft. They should want to wear a pair of heavy leather gloves for this. The string can be pulled really fast and cause a burn or cut on the fingers. Start the clock on the stopwatch and start the motor at the same time. When the string is fully wrapped around the shaft of the motor, stop the clock on the stopwatch and kill the power to the motor. The rpm of the motor is equal to the length of the string in feet, and then divided by the circumference of the motor shaft in feet, divided by the amount of time in minutes that it took to completely wind up. This gets a little confusing because of all the unit conversions. Plug your values into the following formula and you should come out right:

$$\text{rpm} = (L / (3.14 \times D / 12)) / (T / 60)$$

L = Length of the string in feet
D = Diameter of the motor shaft in inches
T = Amount of time for windup in seconds

For example: You have a 15-foot piece of string attached to the motor shaft. The motor shaft has a 0.5-inch diameter. The time for windup was 17 seconds.

$$\text{rpm} = (15 / (3.14 \times 0.5 / 12)) / (17 / 60)$$

$$\text{rpm} = (15 / (1.57 / 12)) / 0.28$$

$$\text{rpm} = (15 / 0.13) / 0.28$$

$$\text{rpm} = 115.39 / 0.28$$

$$\text{rpm} = 412.11$$

In this example you get about 412 rpm.

Will This Motor Work?

Now that we have the specs for the motor, we need toknow wheter it will be strong enough to move our robot. For example, say we have two 12-volt, 100 in-lb stall torque, 300-rpm, 40-amp stall windshield wiper motors. We are building a 58-pound robot. Lots of builders plan on enough torque to carry twice the robot's weight. This means that each motor must be capable of moving the entire 58 pounds by itself. At first glance, you would expect that a motor with 100 in-lb of torque could handle the 58 pounds. The answer is maybe.

To figure out which wheels to use to do, an explanation of torque is required. Torque is the amount of strength the motor has to move something. It is measured, in this book, in units of pound-inches. This means the motor can lift X pounds 1 inch away from the center of the shaft. If a string with a 95-pound weight at the end were attached to a 1-inch lever attached to the shaft, the motor would be able to slowly pick up that 95 pounds. It would be able to hold 100 pounds up in the air while stalled. If you put the string on a 2-inch lever, the motor would be able to hold up only 50 pounds (Figure 6.5).

In Figure 6.6, imagine the *radius* of the wheel as the lever because the wheel contacts the ground at one point. The ground acts as the string and the weight of the robot is the weight in the example. If we use a 4-inch diameter wheel, that's a 2-inch radius. Divide the torque rating by the radius in inches and you are left with pounds. In this case, each motor using 4-inch wheels is capable of carrying 50 pounds. Add the two motors together and they can carry 100 pounds.

This may be a little weaker than we want, but it should work. If it is too weak for your taste, you have a couple choices. You can change motors, change wheels, or use another stage of gear reduction. In this case, we'll go to a 3-inch wheel. This gives us a 1.5-inch radius, and each motor could carry 66.667 pounds, or 133.33 pounds total—well within our plans for carrying twice the weight limit.

Wheel size not only changes the amount of weight our bot can push around but also affects the speed. To figure out at what speed the bot will run, you have to know the circumference of the wheels and the rpm of the motors. Circumference is the distance around the outside edge of the wheel or the distance the wheel covers in one revolution. There are a couple ways to figure out the distance around a wheel. First the hard way: take a piece of string and wrap it around the wheel and measure the length of the string that touches the wheel. That's not really difficult but you should get used to using a formula to figure out

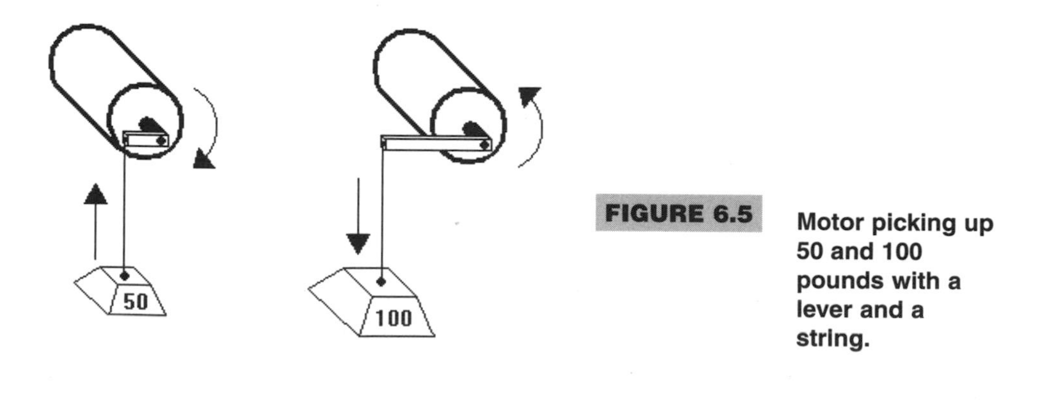

FIGURE 6.5 **Motor picking up 50 and 100 pounds with a lever and a string.**

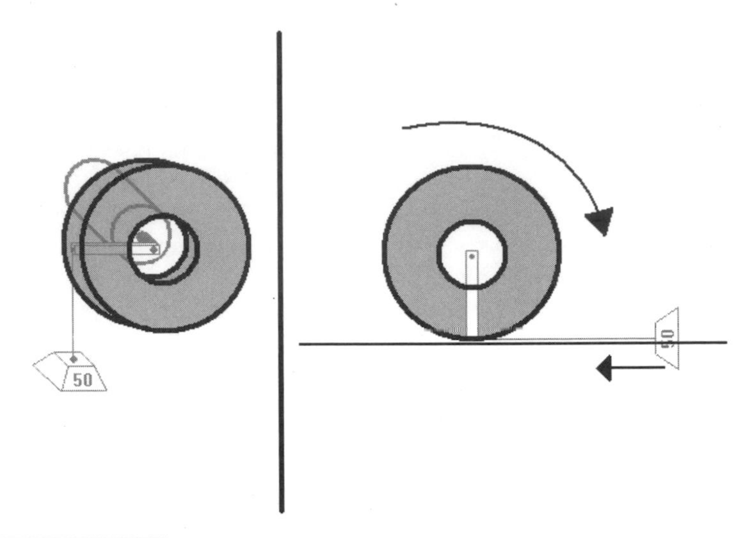

FIGURE 6.6 The radius of the wheel is the same as the lever in previous figure.

things like this. You may not always have the wheel in your possession when you need to know the answer.

Now the easy way—circumference (C) is equal to π (3.14) times the diameter of the wheel (d):

$$C = \pi \times d$$

In this case, the answer is 9.42 inches. The selected motors run at 300 rpm. Multiply 300 rpm by the distance traveled and you have inches traveled in one minute. $9.42 \times 300 = 2826$ inches per minute. Divide by 12 to get 235.5 feet per minute. Divide by 60 to get 3.925 feet per second.

This is rather slow compared to most combat robots. If we go through the calculations for the original 4-inch diameter wheels, we see that our bot moves at 5.23 feet per second, which is still a bit slow. So, the selected motors are a little weak, too slow overall, and changing gearing enough to make them stronger will only make them slower. It would be best to find a motor with the same torque but a faster rpm or with more torque and the same rpm.

LARGER WHEELS

You can sacrifice strength to gain speed by using bigger wheels, or you can get motors that run faster. In Figure 6.7 you see two wheels: pick a spot on each wheel and roll it on the ground until the spot is in its original position. That is one revolution. Do this for both wheels and measure the distances covered by each. The larger wheel covers the longer distance. When attached to a motor that is spinning at a constant speed, the wheels complete one revolution in the same amount of time. However, the larger wheel covers a longer distance. If you cover a longer distance in the same amount of time, you must be going faster.

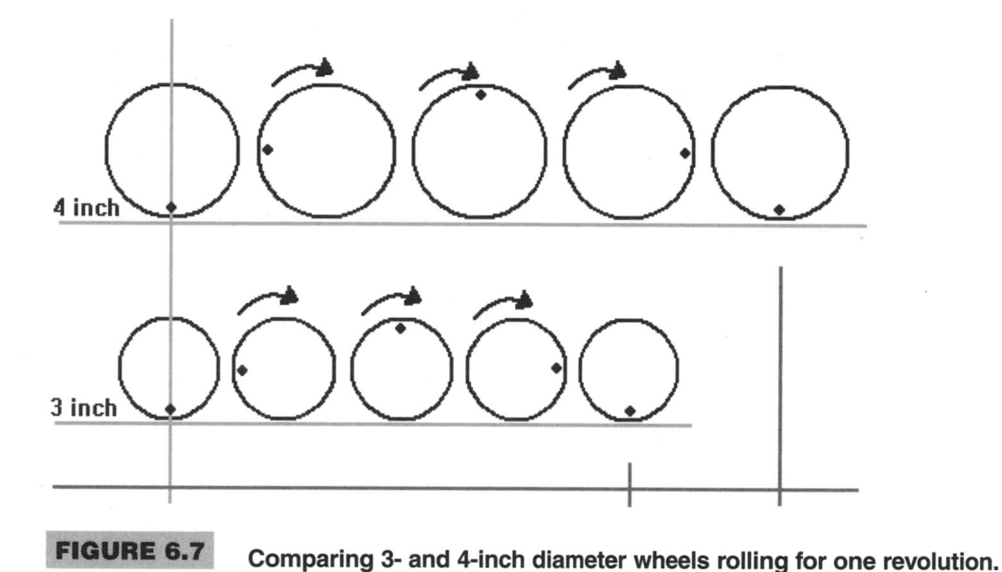

FIGURE 6.7 Comparing 3- and 4-inch diameter wheels rolling for one revolution.

OVERVOLTING A MOTOR

There is one more choice that many veteran bot builders take—you can always take a chance and run the motors at more than the rated voltage. This increases the speed and the strength output. When you double the voltage at which the motor runs, you theoretically double the speed that it turns, double the current that it draws, double the torque that it has, and quadruple the *horsepower* (HP). The large difference in HP occurs because it is calculated by multiplying the rpm and the current draw, which are both initially doubled. Of course, some things limit the actual increase of ratings. The motor may not be physically able to withstand the increase in speed or the increase in current. Wires can burn and magnets can become demagnetized. The heat caused by excess current can melt brush housings or weld brush springs together. In short, you may destroy the motor by using more than the rated voltage. Also keep in mind that you need to recalculate how many batteries you need and whether they can supply the current needed. Also, the speed controllers must be able to handle the increased current draw and voltage.

Motors in Parallel and Series

Like batteries, motors can share loads. When two motors are wired in series and powered by a battery, each motor sees only half the voltage from the battery. If you remember that doubling voltage to a motor doubles its speed and quadruples its horsepower, then it's a small logical step to figure out that cutting the voltage in half will halve a motor's speed and quarter its horsepower. So there's no really good reason to use two motors wired in series to drive the same load. On the other hand, there *is* good reason to wire two motors in parallel. In a parallel configuration, each motor sees the full voltage from the battery. Each motor runs at its rated speed and horsepower. The advantage is the same as having two men rather than one on one side of a tug-o-war rope: more power. Each motor takes over part of the burden of driving the wheel.

However, there is one major drawback to this method. Each motor draws its normal amount of current from the battery, so your batteries must supply twice the current. Your speed controllers must be able to handle that amount of current as well. You see in Figure 6.8 how similar the wiring of two motors to get the desired torque is to the wiring of two batteries to get the desired voltage and capacity.

Nongeared Motors

You can opt to use a motor that hasn't been geared down by the manufacturer, as shown in Figure 6.9. These motors are usually a bit cheaper, and you get to customize more of your drivetrain. The price starts to run back up when you start buying gears, pulleys, belts, and sprockets. For all practical purposes, I've found gears, sprockets, and belts to be interchangeable in designing my robots.

Gears provide several advantages. For one, you can get higher reduction values in a smaller space than you can by using sprockets and chains or pulleys and belts. You can also make right-angle drives with the proper gears. Belts, pulleys, sprockets, and chains have their own advantages. They require a lot less precision when mounting. They allow you to mount a motor in the rear of the bot and drive the wheels in the front. They are also usually cheaper than gears. A disadvantage of sprockets and pulleys is that, because of the size of the sprock-

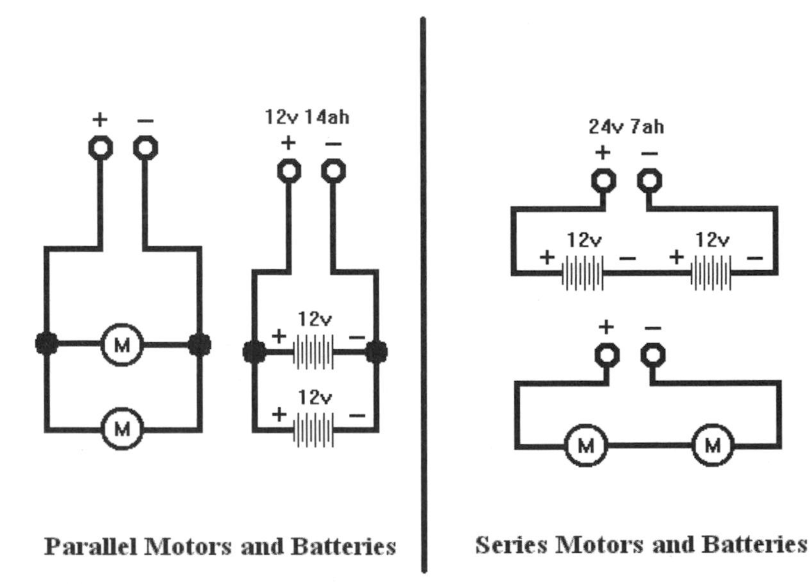

FIGURE 6.8 Wiring motors and batteries in parallel and series.

FIGURE 6.9 **Plain shaft motors.**

ets or pulleys and the number of sprocket teeth that get engaged on the chain, it's hard to get reductions higher than about four or five to one in just one stage. You can do it, but special-order parts are really expensive and the reliability factor is decreased significantly. Also, the longer the chain or belt, the better chance there is that they will come off during the fight. Wheelchair and windshield wiper motors use worm gears to achieve higher ratios. Worm gears can get pretty big, depending on the reduction ratio. Once again, which part you use depends on the design of your bot. The chart of "Proper Gear Usage" in Appendix C gives the relative efficiencies, uses, pros, and cons of the different types of gears.

GEAR REDUCTION

Gear reduction is a straightforward thing and very useful in building combat robots. Figure 6.10 shows a single-stage worm gear reduction. Figure 6.11 shows a dual-stage spur gear reduction. Suppose we have a motor that has 50 lb-in of torque and spins at 2,640 rpm. If you put a 4-inch wheel directly on the shaft, it will spin at 2,640 rpm. From earlier calculations, you see that is about 43 feet per second. That's way too fast to have any real control over it. Now remember that you have to divide the motor torque by the radius of the wheel: you get 25 lb-in per motor. With two motors on a 60-pound robot, the wheels would probably not even move. They would probably melt while your opponent beat on you for 3 to 5 minutes.

FIGURE 6.10 Single-stage worm gear reduction.

FIGURE 6.11 Dual-stage spur gear reduction.

How it works. Since I'm going to talk about gear reduction, I need to explain a little bit about how gears work. I'll talk mostly about spur gears, since the rest of the gear types operate on the same basic principles. The main thing to remember is that any increase in torque results in a decrease in speed. The second thing to remember is that, if you have two or more stages in your gear train, you *multiply* the ratios to get the final reduction value. Do not just add the two ratios together. When a small gear drives a larger gear, the small gear is the *pinion* and the large gear is the *driven* gear. This setup gives you a reduction in speed and an increase in torque based on the number of teeth on each gear. The two gears are in physical contact with each other at all times. Because of the teeth, one gear cannot turn without the other gear moving, too. This is easily demonstrated, as in Figure 6.12, by putting a dot on a tooth of the small gear and a dot on its corresponding valley on the big gear. If the big gear has 20 teeth and the small gear has 10 teeth, you have a 2 to 1 (2:1) *reduction ratio*. If you rotate the small gear (gear A) one revolution, the big gear (gear B) turns half a revolution. If you rotate the small gear another revolution, the big gear completes its one revolution and the dots should match up again. The speed of the small gear (pinion) is twice the speed of the large gear (driven). Hence, the 2:1 reduction ratio. It works the other way, too. If your pinion is larger than your driven gear by the same amounts, you still get a 2:1 ratio. The difference is that you reduce torque and increase speed.

If you secure another pinion gear (gear C, 10 teeth) to the same shaft as the driven gear (B), the second pinion gear (C) rotates at the same speed as the driven gear (B). If you add a second driven gear (gear D, 20 teeth) on a separate shaft, driven by the second pinion gear (C), the second driven gear (D) will turn at half the speed of the second pinion gear (C), or one revolution for every two. Since it takes two full revolutions of the first pinion gear (A) to get the second pinion gear (C) to turn once, it will take four revolutions of the

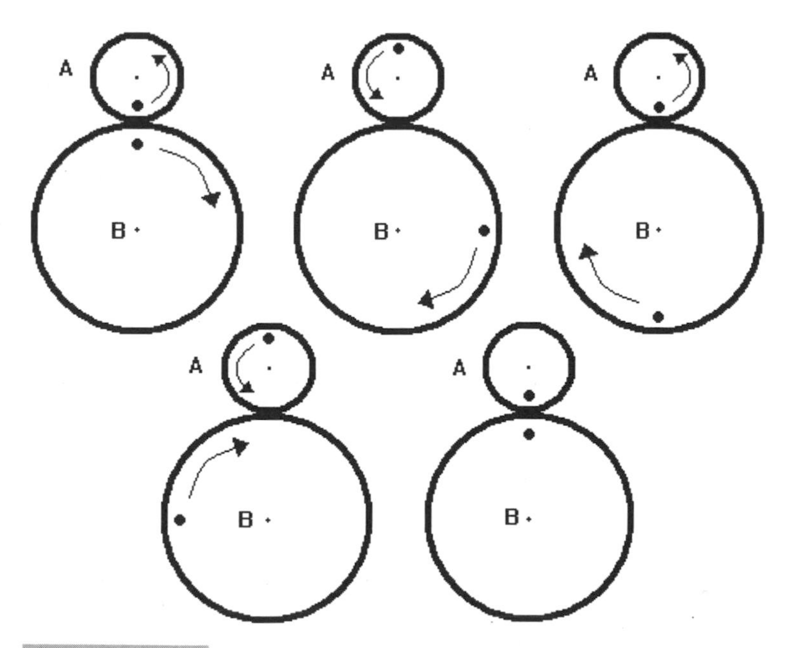

FIGURE 6.12 Pinion and driven gears using dots to show a 2:1 ratio.

first pinion gear (A) to get the second driven gear (D) to turn once. The last gear turns four times slower than the first gear and torque is four times greater. See Figure 6.13. You can continue this until your robot weighs as much as the Space Shuttle and runs slower than a clock but it's best to figure out what gear reduction you need and design to get it.

Each stage of reduction also decreases the efficiency of the gear system because of friction losses and slight misalignments. There are equations that you can use to figure out the losses, but a basic guide is that each stage of reduction loses between 1 and 2 percent. With a two-stage reduction you might see between 96 and 98 percent efficiency but this depends on getting the alignments and center distances pretty accurate. Efficiency also goes down as you use bigger single-stage ratios. We usually do not have to worry about this small amount of inefficiency if we've designed the drivetrain to carry more than the weight of the robot.

Block and tackle. Why gear reduction increases torque is more easily described by using a block and tackle (pulley system) example. If you have a 200-pound block of steel that you want to lift 10 feet in the air, you must exert at least 200 pounds of force on the block in the direction you want it lifted (up) for 10 feet. Imagine that you tie a rope to the ceiling, run it down through a pulley mounted on the block, back up through another pulley on the ceiling, and finally back to you. Then pull on the rope, lifting the block of steel off the ground 10 feet.

Because the weight of your bot is distributed across however many wheels you have touching the ground, you only need a torque rating of at least half the weight of the robot per wheel in a two-wheeled bot. The same thing applies to the block of steel, except that

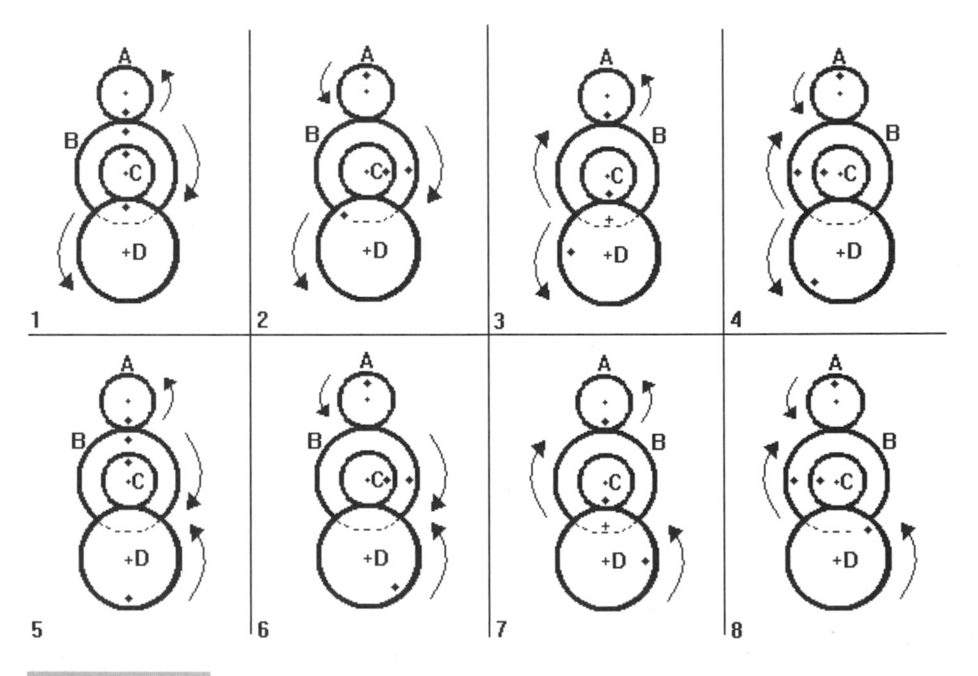

FIGURE 6.13 Pinion and driven gears using dots to show a 4:1 ratio in two stages.

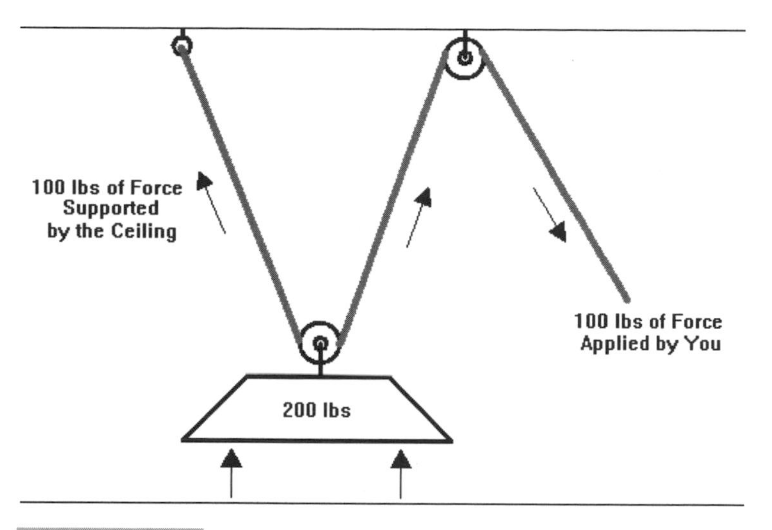

FIGURE 6.14 Weight and pulley example.

you are looking at it upside down. The spot where the rope is tied to the ceiling is holding 100 pounds of distributed weight and the pulley on the ceiling, along with you, is holding the other 100 pounds, as shown in Figure 6.14.

The difference is that now you have to apply only 100 pounds of force but you have to apply it for 20 feet. The same amount of work is completed, just over a longer distance. This tells us that it is possible to trade perceived weight for distance. Gears work the same way. If a nongeared motor is directly attached to the wheel on your robot, the motor must apply whatever force is necessary to move the bot. If you add a gear reduction of 2:1, the motor only needs to apply whatever force is necessary to move half the weight of the bot, but it takes twice as long to do it. Adding another pulley to the block of steel example further decreases the force you have to apply. Adding another gear ratio to the motor example further decreases the force the motor has to apply. It also gives you a multistage gear reduction. Since the motor is going to put out whatever force it's designed to put out, the overall affect is that your bot can carry more than it's own weight.

Gear ratios. Since we want at least twice the bot's weight in torque from the motors, each motor must be able to pull 58 lb-in while driving a 4-inch wheel. Do the math backward to see what the torque is with a 4-inch wheel attached, and you get 25 lb-in per motor again. Through gear reduction, we want to increase the 25 to roughly 60 per motor. Just remember that reducing speed increases torque. To find the reduction ratio, divide the desired torque (60) by the original torque (25) to get 2.4. For every 2.4 teeth on the driven gear, you have 1 tooth on the pinion. The lowest number of teeth I've seen on a gear is 9 or 10, but that's a tiny gear. The same number goes for sprockets, too. So if the pinion has 10 teeth, then the driven gear must have 24 teeth to get a 2.4 reduction ratio. This is not an unreasonable request for a set of gears, sprockets, or pulleys. Now, if you start at the beginning and do the math you'll see that using the original 50 lb-in, multiplied by the gear ratio of 2.4 equals 120 lb-in per motor. This is roughly twice the bot's weight.

As I mentioned before, a gear with 10 teeth is a tiny gear. To get the right size, you must know how much torque the teeth can handle and the size of shaft on which the gear will be secured. If you are using a 1.5-inch shaft, you will never get a 10-tooth, 12-pitch gear on it. Using a manufacturer's catalog or Web site, find the pitch of the gear teeth that can handle the torque your motors are going to produce. Then make sure the maximum bore of the gear with that pitch will fit on the shaft you plan to use. To keep it simple, find the gear with the right pitch and possible bore diameter with the smallest number of teeth (N1). That's your pinion gear. Then multiply N1 by the reduction ratio (2.4). This gives you the number of teeth (N2) you need on the driven gear. If a gear doesn't exist with N2, or N2 isn't a whole number, get the next higher or next lower number of teeth available. Which one depends on whether you want to give up a little strength or give up a little speed.

N2 = Number of driven gear teeth

N1 = Number of pinion gear teeth

r = Reduction ratio

$N2 = N1 \times r$

Now, all that's left is to find the speed. The motor spins at 2,640 rpm. You divide this by the reduction ratio (2.4) and get 1,100 rpm. You take the circumference of the wheel (12.5 inches) and multiply that by the reduced rpm (1,100) to get inches traveled per minute. You divide that by 60 to get inches traveled per second. You divide that by 12 to get feet traveled per second. It comes out to be about 19 feet per second. This is a little fast but still controllable. You may opt to increase the ratio a bit or decrease the size of the wheels a bit to slow it down and give it a little more pushing power.

Suppose you wanted to drive a 110-pound robot with the same motor and the 2.4 reduction gearbox you already built. The robot is going to be underpowered, so you need to increase the gear reduction by adding a second stage reduction. To get the total gear reduction ratio across two stages, you multiply the first stage ratio by the second stage ratio. So, if we want to figure out what it takes to get that one motor to push 110 pounds, you start as though it's only a single stage reduction. Using the same 4-inch wheels, you want to increase the 25 lb-in to roughly 110 lb-in. This takes a 4.4 reduction ratio (4.4:1). So, with a 10-toothed pinion gear, you need a 44-toothed driven gear. Again, not too unreasonable, but we already have 2.4 of the 4.4 ratio and we want to use it. Because the total ratio (T) is the first stage ratio times the second stage ratio (R1 × R2), divide the total by the first stage to find the second stage:

$T = R1 \times R2$

T = Total reduction ratio

R1 = 1st stage reduction ratio

R2 = 2nd stage reduction ratio

Rearranged:

$R2 = T / R1$

$R2 = 4.4 / 2.4$

$R2 = 1.8$

So, the second stage reduction ratio is 1.8:1. This means that with 10 teeth on the pinion you need 18 teeth on the driven gear. Keep in mind that the first stage driven gear and the second stage pinion must be mounted on the same shaft. Not only that, but they must be attached solidly so that both gears turn at the same time. Now that we have a 4.4 reduction ratio we can figure out how fast this 110-pound robot is:

$rpm1 = 2640 \quad r = 4.4$

$rpm2 = rpm1 / r$

$rpm2 = 2640 / 4.4$

$rpm2 = 600$

rpm1 = Initial motor revolutions per minute.
rpm2 = Motor revolutions per minute after gear reduction.
r = Total reduction ratio.

The motor spins at 2,640 rpm. Divide this by the reduction ratio to get 600 rpm. Next, take the circumference of the wheel (12.56 inches) and multiply that by the rpm (600) to get inches traveled per minute. Divide that by 60 to get inches traveled per second. Divide that by 12 to get feet traveled per second. It comes out to be about 10.5 feet per second—a very respectable and controllable speed for a bot.

IPM = Inches per minute   IPS = Inches per second   FPS = Feet per second

$IPM = C \times (rpm1 / r)$ or $IPM = C \times RPM2$

$IPM = 12.56 \times (2640 / 4.4)$ or $IPM = 12.56 \times 600$

$IPM = 12.56 \times 600$ or $IPM = 7536$

$IPM = 7536$

$IPS = IPM / 60$ and $FPS = IPS / 12$ or
$FPS = (IPM / 60) / 12$

$FPS = (IPM / 60) / 12$

FPS = (7536 / 60) / 12

FPS = 125.6 / 12

FPS = 10.466

Sprockets and pulleys. Sprockets and chains work the same way as gears except that they can transmit power across longer distances. The gear tooth ratio becomes a sprocket tooth ratio. You can also do multiple-stage chain and belt reductions, as shown in Figure 6.15. This can get really heavy and complicated, although in complex robots it may be worth the effort.

Belts and pulleys work much the same way as gears and sprockets except the ratio is found by using the *circumferences* of the pulleys in place of the number of teeth on the gear or sprocket. You can use this method with gears and sprockets, but you must use the *pitch diameter* to find the circumference instead of the diameter on the outside edge of the teeth. It's just easier to count teeth on a gear or sprocket. If I use pulleys for a drive system, I like to use *toothed* or *timing belts* and pulleys. They transmit the torque much more efficiently than the v-belt type you can get off lawnmowers.

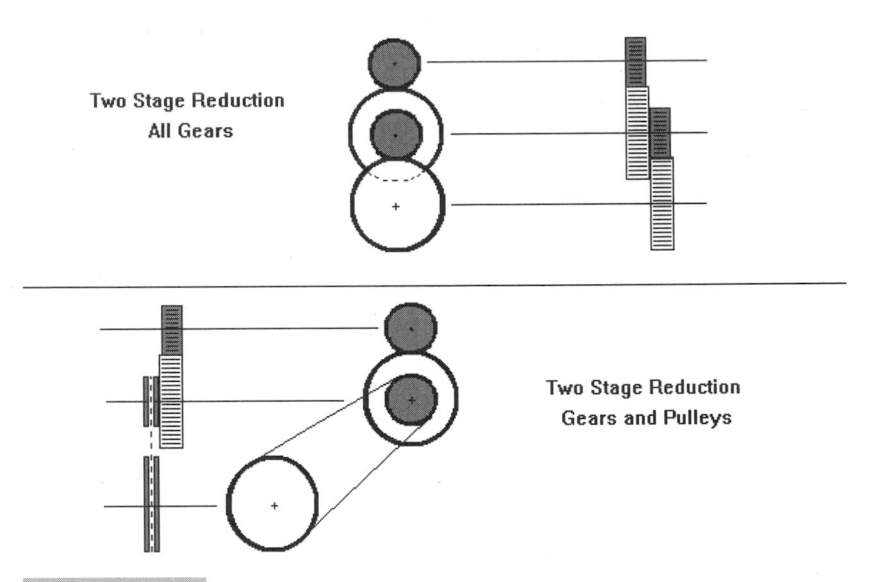

FIGURE 6.15 **Two-stage gear reductions using gears and pulleys.**

Summary

Motors are the muscles of most robots. To use them effectively you must know some facts about motors. Torque, current draw, and rpm are the most useful specs and can help you determine whether the motor is suitable to power your bot. Several tricks can make them more powerful. Gears, sprockets, or pulleys change the speed and the strength of motors. Increasing the voltage does the same thing, but the speed increase may come at a price.

In Chapter 7, we see how to figure out which gears, sprockets, or pulleys we need to get the increases we want. We learn about single- and dual-stage gear reductions. We also go over the terms and measurements you'll need so that you will know what to ask for at the parts counter.

7

GEARS, SPROCKETS, AND PULLEYS

Finding and ordering gears, sprockets, and pulleys is not too complicated. Most towns have a general transmission supply or industrial supplies house that stocks a lot of what you will need and will be able to special-order everything else. You just need to know how to order them. The best thing to do is obtain a manufacturer's catalog for gears, sprockets, and pulleys. I have a Martin catalog that I use—it comes in handy because my local distributor carries Martin products. You can get catalogs from the distributors or by contacting the manufacturer over the phone or by email. With the catalog in hand while designing your drive train, you will know exactly what is readily available and what part number to ask for when ordering it. Some manufacturers spend a lot of money on their catalogs and do not give them freely or do not give them to individuals at all. (They only give them to companies that spend a good bit of money.) Many of these same manufacturers realize that there are small companies or individuals that need parts and they have put up web-based catalogs. Most of these are excellent. If the manufacturer doesn't have one and you can't talk them into giving you a catalog, find another manufacturer. It is their loss. Appendix E contains several manufacturer names that have informative catalogs.

Terminology

Several terms used in the figures must be understood so that you can place an order for gears, sprockets, or pulleys. Other terms are just good to know because you will need them when building your bot. Taking gears as an example, I'll explain what each measurement is that you will need to know while ordering.

BORE

First, you need to know the *bore* of the gear. The bore is the size of the hole going through the center of the gear. It's where the shaft goes. Find the diameter of the shaft. Sometimes there is a *keyway* (slot) cut into the shaft and the gear, running the length of the bore. A *key* is a hardened piece of metal that sits inside the keyway. Keys keep the gear from spinning on the shaft. They are usually used in conjunction with set screws. For simplicity's sake, Figure 7.1 does not show the hub of the gear or the set screw holding the key stock, but it should give you a good idea of how a key works in this application.

Figure 7.1 shows a shaft and gear with keyways cut into them. The slot is the right size to accommodate the key stock (rectangular piece). The key stock is placed in the keyway and the shaft is placed inside the gear. This provides a positive force that keeps the gear from spinning on the shaft. Sprockets and pulleys can use this same method. The *set screw* is normally in the hub of the gear. A set screw is nothing more than a small bolt with a six-sided hole in the end to accommodate an Allen wrench. The set screw is screwed into the hub of the gear and presses onto the key stock. The spinning force is brought to bear on the key stock, and the gear does not spin on the shaft. However, it doesn't take much force to push the gear farther onto the shaft, putting it out of alignment. Set screws should not be depended on to hold gears in place on the shaft. *Shaft collars* provide even more security. Or you can use a lathe to cut a shoulder into the shaft. A spring clip fitted into a groove in the shaft provides additional security. See Chapter 8 for more information on shaft collars.

Most gears come from the manufacturer with a standard bore and no keyway. In the catalogs, you will see a maximum bore category. This tells you the biggest hole you should put in the part. You can order the gears with a special bore and with keyways, but I've found that

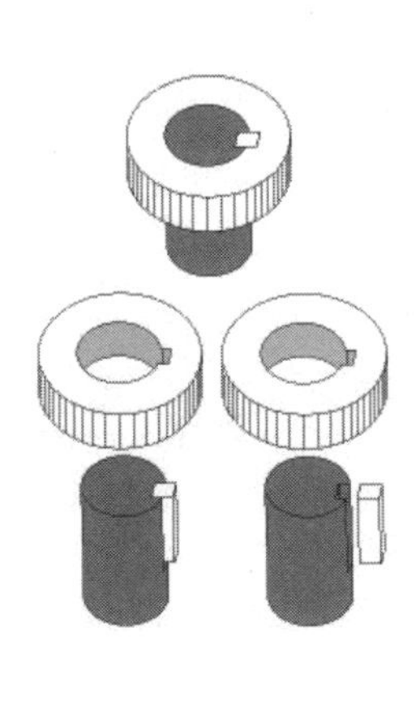

FIGURE 7.1 **Gear, shaft, and key stock installation.**

it can double the price. It's best to learn how to use a lathe and bore these holes yourself. Roll pins can replace keyways to some extent and can be installed using a drill press and a hammer. If you use roll pins, be sure to take measures to ensure that the pins cannot fall out of whatever they are holding. Roll pins also must be able to withstand the torque that the gear will be transmitting. In many cases, they need to withstand more than that because of the forces experienced during a match. In general, roll pins are a bad idea. However, if you find that you must use them, you can add to their strength by inserting a smaller roll pin inside the original pin. Several other types of pins can be used, including solid pins, but none of them is as effective and reliable as a set screw and keyway combination. Squared shafts and bores can be even more reliable if enough material is used for the shaft. Small shafts with odd ends can shear off fairly easily. The larger the shaft, the better.

PITCH

Simply put, the *pitch* represents how many teeth you can fit inside 1 inch around the gear. It is also used to describe the distance from a spot on one tooth to the same spot on the next tooth. On a sprocket, the pitch corresponds to the length of one chain link. Sprockets are not usually described using the pitch. They are described using a standardized number starting at #25 (0.25-inch pitch) and going to #240 (3-inch pitch). Sprockets also come with double and triple rows of teeth to handle increased torque loads. Gears, sprockets, and pulleys also come in metric sizes. If you feel more comfortable using the metric system, feel free to convert to it. Whatever form of device you use, you need a suitable pitch for the torque you are applying. The pitch size mainly depends on the amount of power you are putting into the system, but it also depends on the amount of space you have in your bot. Smaller pitch sizes have smaller components. You can check the manufacturer's specs to find out if the sprocket, gear, or pulley that you have in mind will be able to handle the power you are using. However, many builders run components at several times their rated capacity simply because they have shorter runs. The specs in the catalogs are for industrial-type equipment and give the ratings at which the part will last for a long time while running constantly, that is, 24 hours a day. In many cases, we are only concerned with a part working for several 3-minute matches in one weekend. (In the smaller bots that I've built, I've used #25 and #35 chain and sprockets. I've used 20- and 16-pitch gears and #35 sprockets in bots weighing up to 220 pounds. The smaller, 20-pitch gears wore a bit faster than I preferred, but they held together for nine fights and hours of test run time with no problems before I retired them. I had no problems with the #35 chains and sprockets on the 210-pound fighters. Once, I used a #40 chain and sprocket along with a 12-pitch worm gear on a 325 pound bot. The chain and sprockets held up well, but the teeth on the worm gear were all sheered off during a fight. There were several factors leading to this, the least of which was the fact that it was a 12-pitch. Given that, I'm going to go bigger on the next one.)

NUMBER OF TEETH

Along with bore and pitch, you need to know how many teeth you want on each gear, sprocket, or pulley you order. You also need to know what type of material you want them made from. Many of these devices come in a few different materials, including aluminum, cast iron, steel, and stainless steel. You can get exotic materials, but they usually aren't necessary, are usually hard to find, and are always really expensive. It will save you time if you decide where you are going to purchase your gears in advance and find out which manufacturer the supplier prefers. Then you can get that manufacturer's catalog and have the part numbers of the gears, sprockets, and pulleys you want to order. This shows the company that you know what you are doing—that and talking about your project might make you new friends and possibly get you discounts or sponsorships. Standard sprockets do not come in all sizes and with all numbers of teeth. The "Outside Diameter, Number of Teeth, and Pitch of Standard Sprockets" chart in Appendix C shows you which sprockets actually exist.

OTHER DESIGNATIONS

Other things you must know while designing your drivetrain, especially two- or three-stage reductions, include the *pitch diameter* (PD), *outside diameter* (OD), *mounting distance* (MD), *length through bore* (LTB), *face* (F), *hub projection* (HP), and *hub diameter* (HD) of the gears, sprockets, or pulleys. You can find most, if not all, of these designations in Figures 7.2 through 7.6.

PD is the diameter of the pitch circle. On gears and sprockets, pitch circles run somewhere near the middle of the height of the tooth, where the contact with the mating gear or sprocket is made. On a pulley, it's a bit different. The tension is focused mainly in the belt itself, so the pitch circle is measured on the belt pitch line or at some point in the belt above the actual face of the pulley.

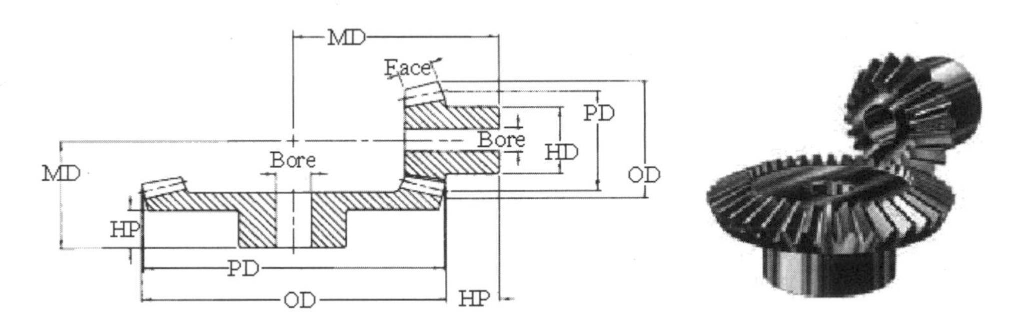

FIGURE 7.2 **Martin spur gear and designations. (Courtesy of Martin Gear Inc.)**

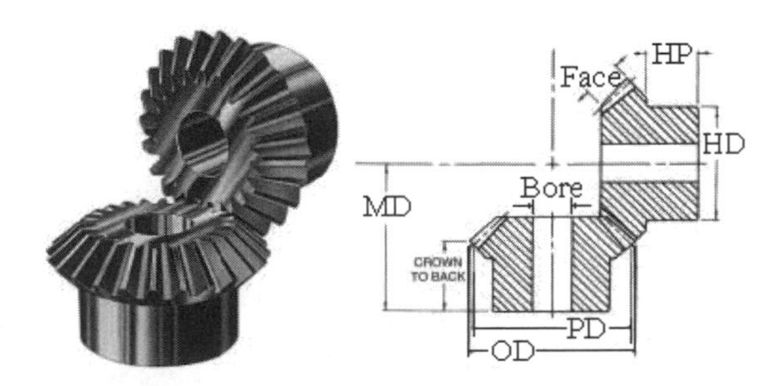

FIGURE 7.3 **Martin bevel gear and designations. (Courtesy of Martin Gear Inc.)**

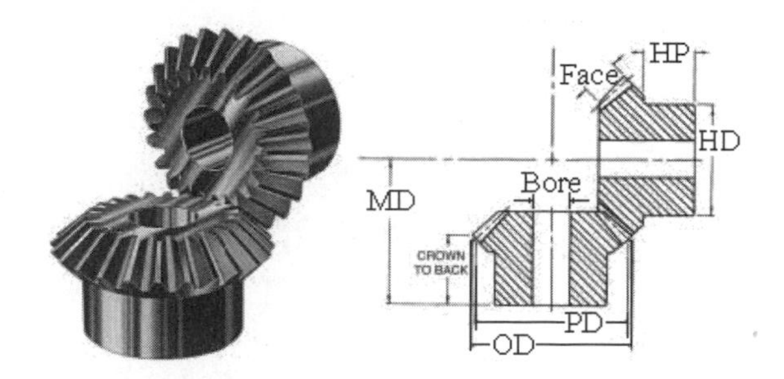

FIGURE 7.4 **Martin miter gear and designations. (Courtesy of Martin Gear Inc.)**

OD is the diameter of a circle that is drawn around the outside points of the part itself. Gears and sprockets have an OD that is greater than their PD. Since the PD of a pulley is measured differently, the pulley's OD is smaller than its PD.

The *MD* for spur gears, sprockets or pulleys is the distance between the shafts of the small and large gears, sprockets, or pulleys. To find that distance for gears, take half of the PD of the small gear and add it to half of the PD of the large gear.

The sprocket and pulley MD is usually calculated when designing the machine, but it does not have to be. In Chapter 21, where I describe building Dagoth, we use tensioners to adjust the slack in the chain. Their corresponding measurement is chain or belt length. To find these lengths use the following formula:

$$L = 2CD + ((\pi \times D1) / 2) + ((\pi \times D2) / 2)$$

where:

D1 = Diameter of the large sprocket or pulley.
D2 = Diameter of the small sprocket or pulley.
CD = Distance between the centers of the two shafts (center distance).
L = Length of the chain or belt.

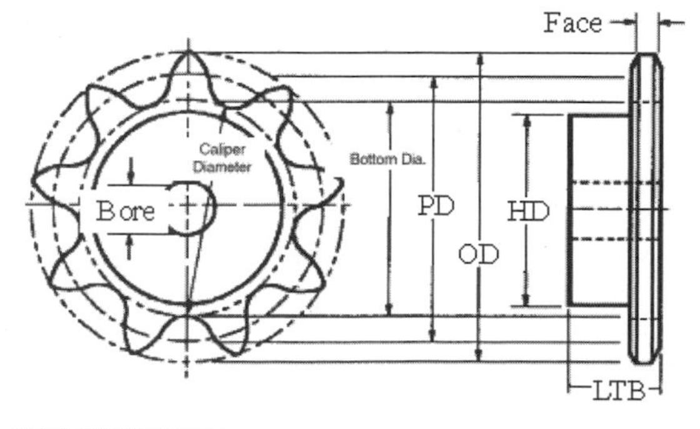

FIGURE 7.5 **Martin sprocket and designations. (Courtesy of Martin Gear Inc.)**

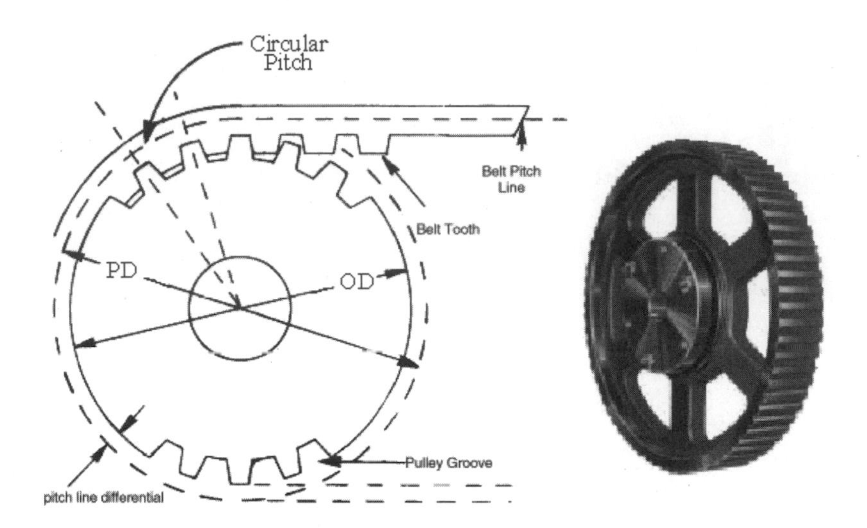

FIGURE 7.6 Martin timing pulley and designations. (Courtesy of Martin Gear Inc.)

If you are using a 1:1 ratio, where both sprockets or pulleys are the same diameter, you can use the following simpler formula:

$$L = 2CD + (\pi \times D)$$

where:
D = Diameter of both sprockets or pulleys.
CD = Distance between the centers of the two shafts (center distance).
L = Length of the chain or belt.

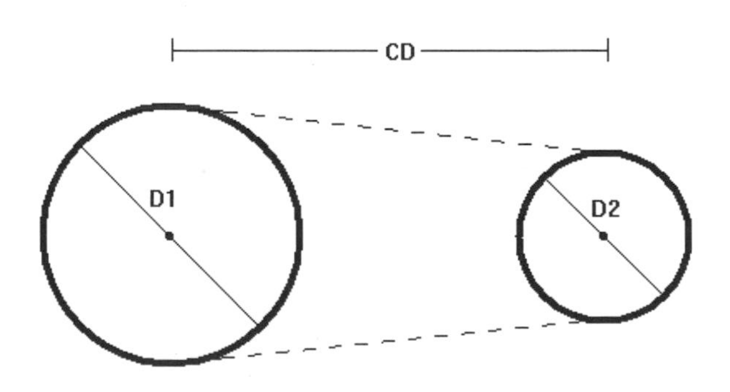

$$L = 2CD + \frac{3.14 * D1}{2} + \frac{3.14 * D2}{2}$$

FIGURE 7.7 Basic belt and chain length formula.

The MD for miter gears is the distance between the center of the bore of a gear and the bottom of the hub of the other gear. To find that distance, take half the OD of the gear and add it to the "crown to back" distance. For the miter gears to work properly, there should be no difference between them. The MD for bevel gears is found in the same manner as for miter gears; the important difference is that the MD for the big bevel gear is larger than the MD for the small bevel gear.

The *LTB* is simply the length of the bore of the part. It is measured by taking the length of the cylinder formed by the bore through the material. There are two important aspects of LTB. First, you must know the LTB so that you can get the correct length of key stock material. Second, you must know the LTB when dealing with plate sprockets that do not have hubs. For these, you must fabricate or purchase hubs to fit to the sprocket. Therefore, plate sprockets effectively have no LTB, HP, or HD. When making hubs for these plate sprockets, examine the LTB, HD, and maximum bore size of the sprockets that have them; make your hubs to closely resemble the manufactured ones, so that they can handle the planned load.

HP and *HD* are closely related. Both affect how a part can be mounted in the bot. You must allow clearance in the design, especially in multistage reductions. (I once had to return and reorder several gears because I had not taken the HD into account while designing the gearbox. I was lucky that I had not already made the bore to my required size when I noticed the problem.)

Chain Positioning and Lubrication

Sprockets are a little forgiving with respect to mounting and alignment. The main thing to look out for is the tension of the chain. There should be a small amount of slack in the chain, but not so much that it can jump off the sprocket. In any chain and sprocket assembly, one side has more slack than the other. This is the side that the drive sprocket is pulling toward (see Figure 7.8). Pay attention to the direction of spin of the small sprocket. Follow the diagram for the preferred positional mounting of the sprockets, and always keep in mind that you do not want to have the loose side of the chain hanging down toward the tight side. This could possibly get tangled while the frame of your robot is flexing from a big hit during a fight. The only catch to this rule of thumb is that, if you are using a chain to drive the wheels of the bot, then the slack side of the chain changes when the motor changes direction. You can keep this from being a problem by using chain tensioners or by having one axle position adjustable, so that you keep the slack in the chain down to a bare minimum and can adjust it after each match. Also keep from positioning the sprockets in a vertical line with one another. Any slack in the chain is focused around the bottom sprocket and may cause it to jump off. One other thing to keep in mind while mounting is that chains must have tension on them at all times. It is possible to lose tension during a fight and cause the chain to jump off. I once built a bot with a single-stage sprocket reduction. The motors were mounted on a long span of steel support and were the highest part of the bot. It looked solid enough. However, during a fight, the bot was flipped upside down. The impact of landing was enough to flex the motor supports and push them toward the driven sprocket enough to have the chain come off. There were a couple very simple things I could have done to keep this from happening, but I was new to the sport and didn't have the experience to see them before it happened. The moral of this is that you always need to mount drive components rigidly.

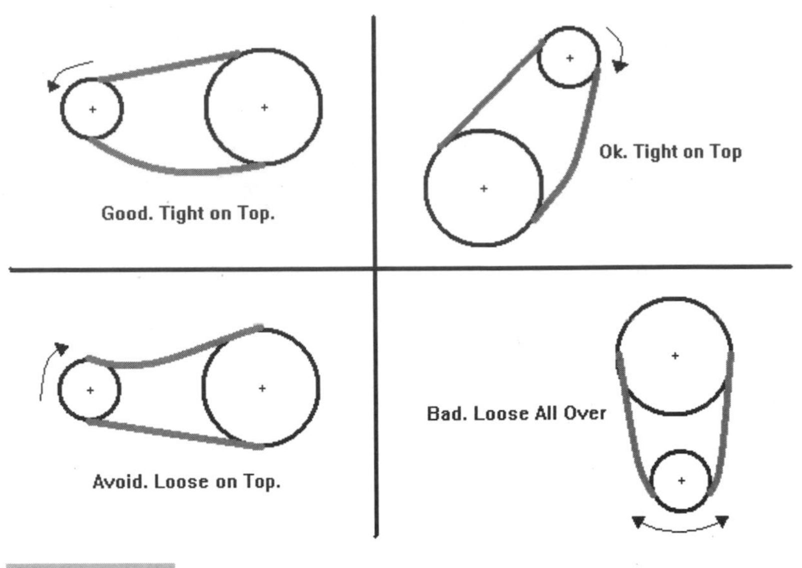

FIGURE 7.8 Chain positioning.

Lubricating chains, sprockets, and gears is really simple. You can always squirt some oil on everything, but this is really messy and can cause problems with your bot's electronics if they aren't protected from splatter. Really heavy oils can be used, but most may fling off chains or gears when they are running at high speeds. Grease is not a whole lot better than oil; it hangs onto the components better, but you still get some splatter. Both grease and oil tend to pick up small metal shavings and dirt. Everyone knows that one small but solid piece of metal can cause a set of gears to lock up and can cause a chain to derail. Graphite powder is the best thing I've found to lubricate my bot's drive train. (You still need to have your electronics in an enclosure for their protection because graphite is a conductive material.)

(Some more exotic chemical mixtures on the market are similar to grease, but claim not to attract dust and dirt, but I haven't tried them yet. Try whatever you can get your hands on and see for yourself which one is best.)

Summary

In this chapter, you learned the terminology of gears, sprockets, and pulleys, which is half the battle in designing drive systems. Combined with the knowledge you gained in Chapter 6, you should be able to put together some simple chain or gear drivetrains. You should at least be able to walk into a parts supplier and sound like you know what you are talking about.

In Chapter 8, you will learn about the axles, bearings, and bushings that allow your drive train and weapon systems to run smoothly and stay together. Because of weight or speed restrictions, you must be able to choose the correct components, and you'll find out how to do that next.

8

AXLES, BEARINGS, AND BUSHINGS

Different types of forces are produced in a bot drivetrain, and axles and bearings must be designed to handle each type of force. Many types of each are available, so the job is not too difficult.

Axles or Shafts

Axles or shafts are an integral part of your drivetrain and sometimes your weapon systems. You mount wheels and gears, sprockets, or pulleys on them. They need to be strong and supported so that they do not bend. They support the weight of your bot (and possibly the weight of your opponent's bot). Axles or shafts are rated by the maximum torque they can handle, and sometimes by a limiting or maximum rpm. Maximum torque can be used to determine hp (or vice versa) by the following formulas:

$$\text{hp} = \frac{\text{Torque (in.-lb.)} \times \text{rpm}}{63{,}000}$$

$$\text{Torque (in.-lb.)} = \frac{\text{hp} \times 63{,}000}{\text{rpm}}$$

These formulas give you the theoretical maximums based on industrial criteria. However, bot builders tend to ignore some industrial maximums for shafts and motors for the simple fact they are derived through focusing on much different goals. Industries rely on machines that often need to operate for 24 hours, 7 days a week. Bot builders require operation for about a week at the most. Because of this, there are a couple rules of thumb that I have used successfully in my bots. Just remember that rules of thumb do not replace the maximum rpm of a part—if you spin a saw blade beyond its maximum rpm, you risk destroying it. Pay attention to the manufacturer's specifications, but realize why they were created.

In the higher weight classes, it is my experience that the material you use for axles is not as important as the size of the axle. I have never used an axle smaller than 1-inch diameter in a robot that weighs more than 100 pounds. I also have never used an axle smaller than 1/2 inch in a bot that weighed more than 50 pounds. And finally, I have never had a bent axle. On spinner weapons in bots weighing more than 100 pounds I try to use shafts that are 1-1/2 inches in diameter or more. If you run into weight limit concerns, you can almost always bore out the axle or shaft to lose a little weight. A rule of thumb that I observe is to never bore out more than what would leave a 3/8-inch shaft wall. That rule changes depending on the size of the shaft and its role in the mechanism. If I am using a large shaft for a spinner weapon, I try to leave 1/2-inch walls. By following this rule of thumb, I have been able to use only mild steel shafts with no extra hardening.

Some people like to experiment with different hard materials. Drill rod is a popular axle material, although it is brittle. Chrome-moly is another popular shaft material but it requires additional hardening to be effective. Many of the industrial supply houses carry regular steel, drill rod, and chrome-moly shaft materials. They also carry prehardened steel, stainless steel, chrome-plated, and even ceramic shafts. Take your pick according to the money you want to invest and what you are equipped to use. Do some experimenting to figure out what is best for you.

One other aspect of shaft design is how the shafts are mounted. This is probably more important than the material and size of the axle—in fact, I'm sure it is the main reason I have had no bent shafts. To get away with using a smaller diameter material, you can support the shaft on both ends with some type of bearing, as shown in Figure 8.1. The ideal configuration for this has the gear, sprocket, pulley, or wheel mounted on the shaft between the bearings, so that both bearings share the entire load imposed by the drive mechanism. This is called a *supported axle* or shaft.

The other type of axle or shaft is called *overhung*: the shaft has two supporting bearings, but the gear, sprocket, pulley, or wheel is not mounted between them. This is demonstrated most often with bots that have wheels on the outside of their body, as shown in Figure 8.2. The advantage to this is that it is more difficult to lift the bot completely off its wheels. The disadvantage is that you will need larger, sturdier shafts that can handle repeated slams by your opponent without bending. You will also need bearings that can handle more load. Overhung wheels are also more exposed to attack than supported wheels. If you are using pneumatic wheels, this can be particularly deadly.

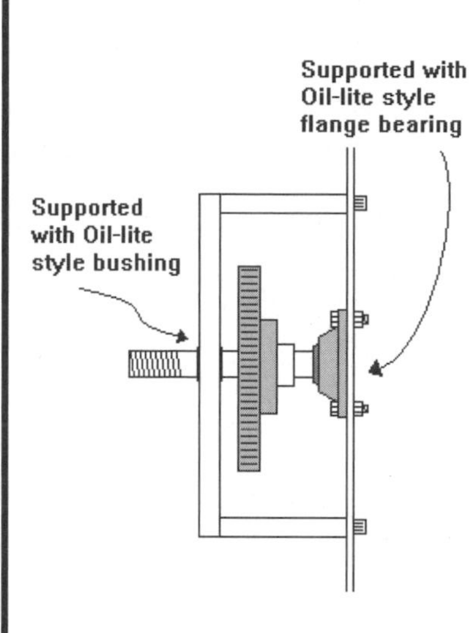

FIGURE 8.1 Supported shaft with gear.

FIGURE 8.2 Bot with overhung wheels.

Shaft Collars

Most of the time, you must restrain the sliding action of an axle. Many large ball bearings have set screws that do this job (see Chapter 7). Some of the nicest bearings are the *cam-lock bearings*. These bearings have an offset hole and a second piece that tightens down onto the shaft via large threads on the main bearing. The second piece has an offset hole, too. When tightened, the holes try to line up and simply tighten down on the shaft. Smaller ball bearings, *UHMW* bearings, and *Oil-Lite* style bearings and bushings do not usually have set screws either. If you use a bearing without a set screw, you will probably need a shaft collar. Shaft collars, in their simplest form, are a ring of material with set screws. Figure 8.3 shows two shaft collars, one on either side of the gear.

Set screws are not always the best idea for holding a shaft in place, because they cause damage to the shaft when clamping down and only apply pressure in one place. A clamp-on style collar is much better. The one-piece clamp-on collar creates pressure around the shaft for a full 360 degrees. They also provide a greater clamping force than the set screw collar without causing damage. The two-piece clamp-on collars do the same job as the one-piece with increased ease of installation—there is no need to take apart the shaft/bearing assembly to install them. Hinged clamp-on collars are similar to the two-pieced collars but can be installed without the risk of dropping half of the collar into the bot. Figure 8.4 shows several types of shaft collar.

Occasionally, you may want to use a threaded shaft in your bot. I have used one in a spinner weapon. The idea was that the spinner shaft could spin freely in the blade when it came in contact with the opponent, to allow the entire amount of energy to be transmitted into the opponent. The gears used to drive the shaft were spared, because the shaft kept

FIGURE 8.3 Shaft collars with set screws.

FIGURE 8.4 **Other types of shaft collars.**

spinning instead of coming to a sudden halt. The problem we had to overcome was how to allow the blade to spin freely on the shaft and still be able to use the shaft to spin the blade. A very simple slip clutch mechanism was designed. The blade rested between two pieces of material that pressed against it. Trial and error were used to determine the amount of pressure on the blade that would allow the shaft to spin the blade and still allow the shaft to spin inside the blade when it hit an opponent. At first, we used a couple of nuts that fit on the threaded shaft to lock it in place. The nuts did not hold exactly where they were supposed to and in most cases tightened further onto the blade. This meant the shaft would not spin when the blade stopped and it put unwanted forces on the gears. We switched to a threaded shaft collar. The collar was screwed down onto the shaft and tightened to apply the correct pressure. Once the correct pressure was reached, the shaft collar was tightened onto the shaft so that it could not move. This held the proper tension on the spinner blade throughout all hits on the opponent.

Ball and Roller Bearings

Bearings and bushings are a big part of making things run smoothly by reducing many different types of friction between parts. The two main types of bearings that I use are *ball bearings* and *roller bearings*. Ball and roller bearings reduce friction between parts by minimizing the contact made between parts. A ball or roller bearing contains many small

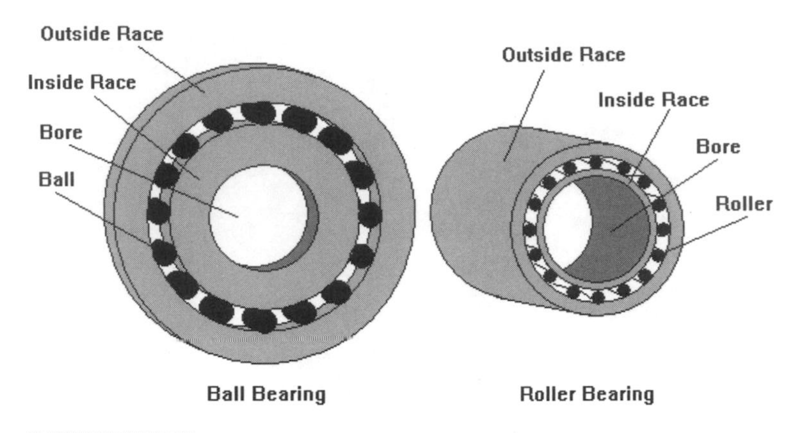

FIGURE 8.5 Ball and roller bearings.

balls or rollers riding between rings, called a *race*. One race is in full contact with the shaft and one race is in full contact with the hole. Because the balls between the rings are spherical and very smooth, the number of contact points between the shaft and hole is minimized drastically. This reduces the amount of friction so much that it appears to almost disappear. Figure 8.5 shows a ball bearing, a roller bearing, and their designations.

Roller bearings are very similar to ball bearings, except that the roller is long instead of spherical. Roller bearings are used mainly in low profile bearings. Use roller bearings when your part does not have enough clearance. Figure 8.6 shows some assorted ball and roller bearings without housings.

FIGURE 8.6 Assorted ball and roller bearings.

Oil-Lite Style Bearings

The Oil-Lite style bearing is one of many types of *plain bearings*. A plain bearing is simply a bearing that doesn't have moving parts (balls or rollers). (These are also known as SAE 841 bronze bearings.) These bearings are made of an *alloy* of tin, carbon, and copper. The pores in the alloy are impregnated with oil. When using these bearings, the friction of the shaft rotating in the dry bearing creates heat. The heat makes the pores open up, and the oil is released. The oil coats the inside of the bearing, forming a blanket of liquid around the shaft and helping to reduce the friction. Figure 8.7 shows some assorted Oil-Lite style bearings in aluminum housings.

Ball and roller bearings are used heavily for high and low rpm devices. Depending on their construction, some ball and roller bearings can handle in excess of 140,000 rpm. However, at that speed, their load capacity is usually less than 100 pounds. The Oil-Lite bearings are good for low rpm devices, even though the calculated maximum rpm for a 1-inch diameter shaft is about 4,500 rpm. Size and cost are a couple reasons to use the Oil-Lite style bearings instead of ball bearings. Oil-Lite style bearings have a much lower profile, can fit in smaller spaces, and cost a bit less than ball bearings. For example, a standard 1-inch inside diameter ball bearing has a 2-inch outside diameter without a housing and can cost more than $8. A standard 1-inch inside diameter Oil-Lite style bearing has

FIGURE 8.7 **Oil-Lite style bearings.**

between 1 1/8-inch and 1 1/2-inch outside diameter and can cost as little as 60¢ from the same company. Several different bore lengths are available in the Oil-Lite style.

LOADING AND RPM

Ball and roller bearings can carry much higher loads and rpms than Oil-Lite style bearings. You must calculate the maximum loads and maximum rpm that an Oil-Lite bearing can handle before you put one in use. Use the following formula and the "Plain Bearing Pmax, Vmax, and PVmax" chart in Appendix C to find the maximum shaft rpm. For inch bearings (non-metric), maximum velocity is stated in surface feet per minute (fpm).

$$\text{rpm} = \frac{\text{Vmax}}{0.262 \times \text{Shaft Diameter}}$$

Vmax is the maximum velocity or speed that a bearing can carry at light loads. Vmax is specified by the manufacturer in its catalog or in Appendix C.

Use the following formula to find the maximum bearing load. For inch bearings (non-metric), maximum load is stated in pounds per square inch (psi):

$$\text{Maximum bearing load} = \text{Pmax} \times \text{Bearing Length (inches)} \times \text{Shaft Diameter (inches)}$$

Pmax is the maximum load a bearing can handle at 0 rpm. Pmax is specified by the manufacturer in its catalog or in Appendix C's "Plain Bearing Pmax, Vmax, and PVmax" chart.

Once you've chosen a bearing based on Vmax and Pmax, use PV as the final check to make sure the bearing can handle the combined load and rpm requirements. If the actual PV is less than PVmax, the bearing should stay together while in use.

$$\textbf{Actual PV} = \left(\frac{\textbf{Actual Load}}{\textbf{Bearing Length} * \textbf{Shaft Diameter}}\right) * (\textbf{RPM} * \textbf{0.262} * \textbf{Shaft Diameter})$$

(As mentioned before, Vmax, Pmax, and PVmax values for plain bearings of several material types are given in Appendix C. I have also listed the recommended shaft hardness and operating temperature ranges there.)

Forces

Several types of bearings are used to minimize the different types of forces that act on rotating shafts. Two forces that act on rotating shafts are radial and thrust. *Radial force* happens, for example, when a wheel rotates on a shaft. The radial force pushes the wheel against the shaft. It changes position around the wheel, depending on which part of the wheel is on the ground. A *thrust force* happens when you push on the wheel, as if trying to push the wheel farther onto the axle. Thrust forces happen when the wheel goes over bumps, too. The best bearing to use if you plan on experiencing both thrust and radial loads is the tapered, roller bearing (Timken bearings). This bearing has rollers, one end larger than the other, that are positioned so that they are at an angle to both the thrust and radial loads. Because it uses rollers, and the forces are spread out over a line instead of a point on a ball, the tapered roller bearing can handle much more, and both kinds of, force without damage. When you are designing your bot, you must pay careful attention to where axles experience force and friction and use the appropriate bearing or combination of bearings.

Manufacturing Style

For all bearings, there are many different manufacturing techniques. Some bearings have a flange made into their housing so that they can be pressed only so far into a piece of work. Other bearings are inserted into housings that are flanged or are base mounted (*pillow blocks*). Most bearings have a few different styles that suit the intended use. *Open bearings* cost a little less and have a slower top speed. Use these if you can be assured that no dirt will get into the balls and races. *Shielded bearings* cost a little more and have a higher top speed. Use these bearings if you are relatively sure that no dirt will enter the bearing. *Sealed bearings* positively keep dirt out, preserve lubricants, and reduce operating noises. They have the same top speed as the shielded bearings and give more benefit for more money.

Summary

In this chapter, I offered some rules of thumb for choosing axle and shaft material. We also saw how using bearings can reduce friction, which is great because it cuts down on motor strain and battery drain.

In Chapter 9, we take a look at pneumatics. I'll give you basic descriptions of major components and show you how those components fit together to form a working pneumatic system. We'll also cover some formulas that will aid you in designing a pneumatic system.

9

PNEUMATICS

Air-powered weapon systems are second in popularity only to electrically powered systems, only because they are more difficult to build and maintain. They can be more dangerous to a builder because of the added complexity and incredible potential velocity of parts of the system. The basic system consists of the air supply, the regulator, the solenoid valve, and the cylinder. The actual system has a burst disk for safety and a gauge or checkpoint to check the gas pressure of the tank. It also has a manual valve that can be used to purge the pressure after the regulator, and a relief valve and another checkpoint after the regulator. (I am not going to instruct you on building an entire system because of the inherent danger of pneumatics. Once you get a feel for the major parts discussed here, get a pneumatics expert to help design and build the system according to the rules of the competition you will attend.)

CO_2 Tanks

The term *air supply* is a little misleading because most people do not really use air. Most people use a *compressed gas* like carbon dioxide (CO_2). This is the same stuff used in paintball guns, fountain drink machines, and welding, so it is readily available to most builders. CO_2 at room temperature is a gas. When you put CO_2 gas under enough pressure, the gas turns into a liquid. Letting the pressure off the tank into your pneumatics system lets some liquid return to gaseous form. The expansion of the liquid into gas and the gas

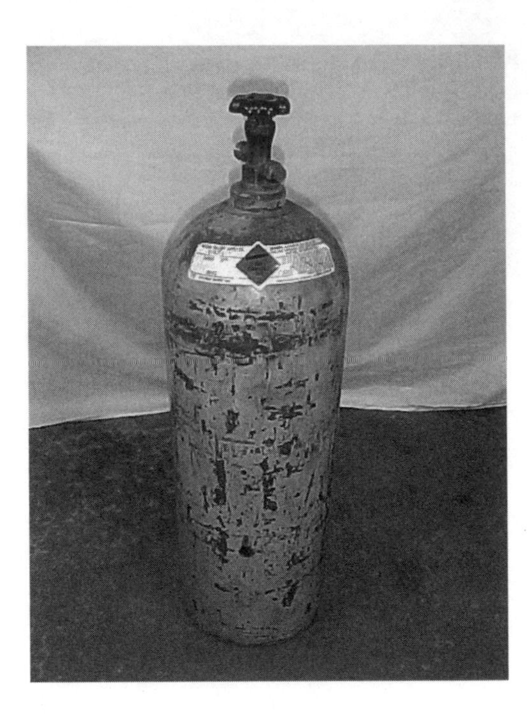

FIGURE 9.1 CO_2 supply tank.

itself push the pneumatic actuator in and out. Every time you command the cylinder to move, more liquid converts to gas until there is no more liquid in the tank. Figure 9.1 shows a large CO_2 tank that I use for testing as well as a welding gas supply.

Take care not to let your CO_2 tank get heated beyond room temperature. This can cause the liquid in the tank to expand into gas and possibly burst the tank. The burst disk should pop before the tank does, but with this type of system there is absolutely no reason to take a chance. A burst disk is, of course, a safety device that is used to help ensure that the tank is not overfilled. You must also make sure the tank is protected from possible puncture. This could cause not only a loss of power to your weapon but also an explosion, depending on the tank.

Paintball tanks are designated in ounces (oz). A 20-oz tank holds 20 ounces of liquid CO_2 by weight. Ask for an anti-siphon tank: when tilted or upside down, regular tanks can allow the liquid CO_2 to enter the regulator and possibly the valves of the system. Regulators and valves do not like that. An anti-siphon tank keeps this to a minimum. When traveling, some airlines do not like paintball or other types of tanks in luggage, and none are going to let you transport a filled CO_2 tank. Before you leave for a competition, check with the locals for a place that can fill your CO_2 tanks.

Regulators

A regulator is shown in Figure 9.2. Liquid CO_2 in the tank is pressurized at about 850 pounds per square inch (psi) depending on temperature. Normal operating pressure for most air cylinders, hoses, and valves is about 150 psi. If you were to release the 850 psi of

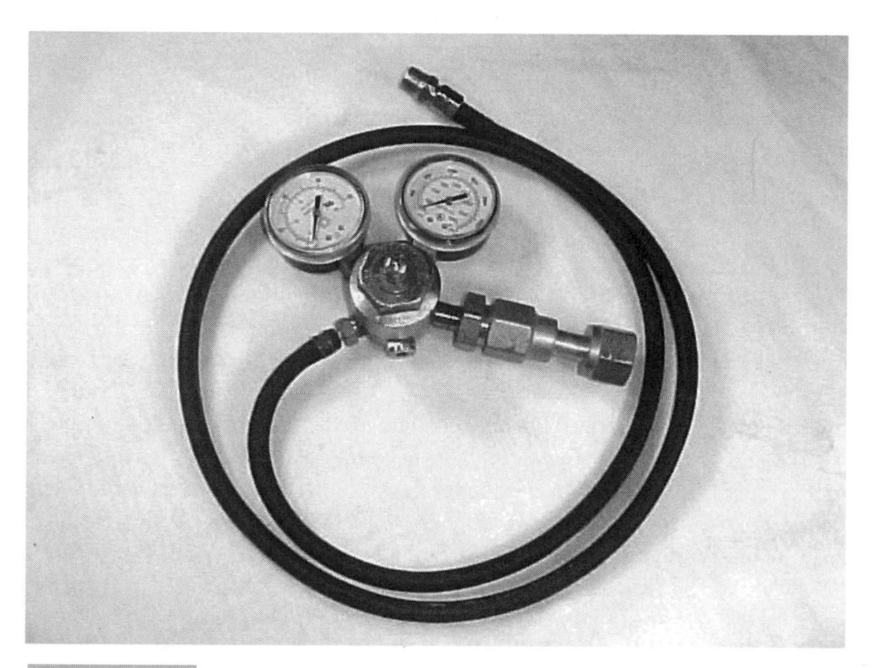

FIGURE 9.2 CO_2 regulator.

the tank into the 150 psi cylinder, the cylinder could explode. It would, at the very least, launch the rod out of the end of the cylinder and through the walls of your workshop. Needless to say, you need something to change the 850 psi of the tank into the standard 150 psi operating pressure of the rest of the system components. That something is called a regulator. Regulators allow only the right amount of pressure into the system. A regulator is usually mounted directly onto the tank, or it can be mounted to a manifold or simply inline. (A *manifold* is basically a chunk of metal with threaded holes. One of the holes is a match for the supply tank; one is a match for the regulator; one is a match for either a gauge or a tap, so that the pressure can be measured easily.)

Buffer Tank

The drawback of using a regulator is that most are not designed for a high flow rate. That means that only so much gas can get through the regulator at one time. This can have a devastating effect on the power of your system if it requires more gas than the regulator can supply in an instant. Flow rate is the key to power. To overcome this problem, many people use a *buffer tank*. This is an empty tank that is fed directly by the regulator and directly feeds the rest of the system. The output of the buffer tank can handle the high flow rates required. So, while you aren't using the weapon, the regulator is dumping gas into the buffer. The buffer holds the gas until the solenoid valve opens, at which point it dumps it all in a split second. Then the regulator starts feeding the buffer tank again. There can be a short delay between full-power weapon uses, but nothing really noticeable. In fact, I've seen systems that could fire nearly as fast as the operator could press the button.

Solenoid Valves

The *solenoid valve* (shown in Figure 9.3) is a really important part of your pneumatics system that controls when the supply gas is released into the cylinder. The solenoid valve is electrically actuated by, you guessed it, a solenoid. Remember that the key to a good pneumatics system is flow rate. The solenoid valves with the best flow rate actually have two valves in one. The solenoid controls a small valve. The small valve controls a small stream of air that operates a large, high flow rate valve. Depending on the type of cylinder you plan on using, you will need a three-port or four-port valve to operate it. (There is also a five-port valve—it is more expensive but offers an even better flow rate.) A three-port valve has one input port, one output port, and an exhaust port to make a cylinder move in one direction; it is used to control a single acting cylinder. A four-port valve has one input port, two output ports, and an exhaust port. The two output ports are connected to a double acting cylinder. One produces the extend motion and the other, the retract motion. The five-port valves have the same ports as the four port valves, plus a second exhaust port. In this configuration, each output port has its own exhaust port, which gives a freer flow to the exhausted gases.

Cylinders

Small and large examples of double acting air cylinders are shown in Figure 9.4. A pneumatic cylinder is a hollow tube with a *piston* inside and sealed ends. The piston is a disk with an airtight seal against the walls of the cylinder and a rod that extends outside the

FIGURE 9.3 **Pneumatic solenoid valves.**

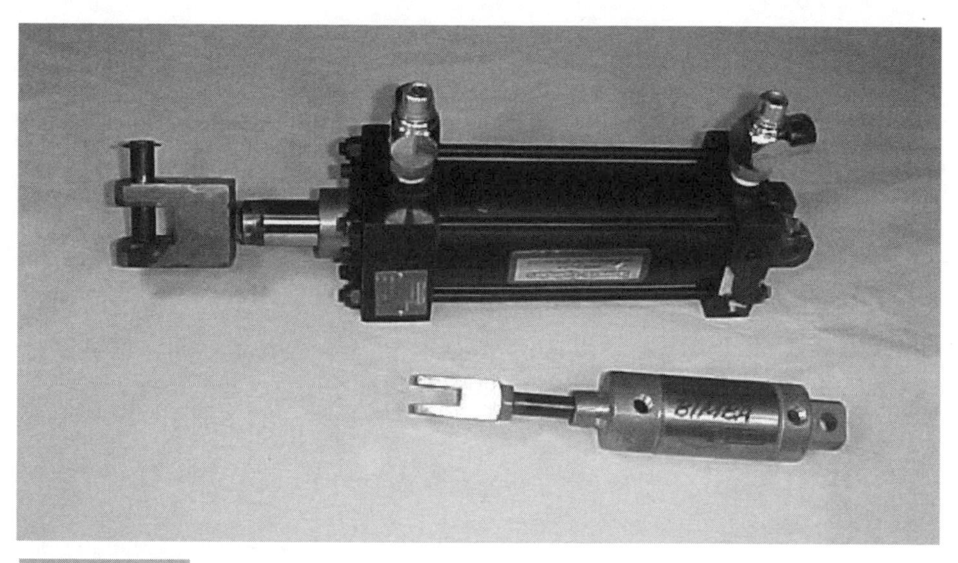

FIGURE 9.4 **Double acting air cylinder.**

cylinder. The rod actuates whatever type of weapon it is attached to, be it a hammer, spike, ram, or lifter. Two types of cylinders are used by builders: *single acting* and *double acting* cylinders. I mentioned them briefly before. The single acting cylinders have one input port. CO_2 enters the port and pushes on the piston to extend it. When the CO_2 is turned off, the piston retracts itself via spring, bungee cord, gravity, or some other method. This type of cylinder is controlled by a three-port solenoid valve.

A double acting cylinder uses CO_2 to extend and to retract. There are two input ports: when CO_2 enters the first port, the piston is extended. When the valve closes that port and CO_2 enters the second port, the piston is retracted. As the valve closes the CO_2 supply to either of the ports, the CO_2 in the cylinder escapes through the exhaust port or ports on the valve.

Bore and *stroke* characterize cylinders. The bore is simply the diameter of the piston head—the larger the diameter, the more pushing power your cylinder has. The stroke is the length of travel of the piston. Thc stroke determines how far your cylinder extends.

Calculations

If you are going to use pneumatics, you must know how to figure out the amounts of gas and the size of the cylinder to use. The cylinder bore and the pressure of your system determine the power of your system.

FORCE

In fact, there is a formula to find the force created by the piston. It is pressure (*pounds per square inch* or psi) of the gas times the area (square inches) of the piston head on which you are pushing. On the rod side, you figure it the same way, only the area of the piston is actually a little less because the rod is taking up space. Normally, a piston is round, so the area of the piston is equal to half the bore squared times π (3.14).

Force = Pressure × Area

Area = $(\text{Bore} / 2)^2 \times \pi$

NUMBER OF SHOTS

You most likely will want to know about how many shots you can get out of one tank of CO_2. That answer is a little cloudy, but is not too hard to get an approximate number if you know something about the properties of gases. Pressure multiplied by volume in one condition equals pressure multiplied by volume in another condition. Also, when the pressure of a gas changes, the volume changes. This is shown in the formula known as *Boyle's law*:

$$P_1V_1 = P_2V_2$$

If you know any three variables, you can find the fourth. For example, a CO_2 gas stored at 850 psi in a 51-cubic inch (cu in) tank forms a 1.7-cubic foot (cu ft) cloud when released into the air. Here is the actual math: P_1 equals the stored pressure of 850 PSI. V_1 equals the 51-cu in volume in the tank. P_2 equals 14.7 psi (regular air pressure). V_2 is the volume we want to find at 14.7 psi.

$$P_1V_1 = P_2V_2$$

$$850 \times 51 = 14.7 \times V_2$$

$$43350 = 14.7 \times V_2$$

$$43350 / 14.7 = V_2$$

$$V_2 = 2948.98 \text{ cu in}$$

Convert it to cubic feet by using the conversion factors in Appendix B.

2948.98 cubic inch divided by 1728 = 1.7 cubic feet

This does not give you the number of shots to be had from your tank, because you are not releasing your CO_2 into the air. You are changing the pressure of your tank (850 psi) to the pressure in your system (150 psi) by running the gas through the regulator. So, do the math again to find out the volume at 150 psi. P_1 equals 850 psi in the tank. V_1 equals 51 cu in in the tank. P_2 equals 150 psi in the rest of the pneumatic system. We want to find V_2:

$$P_1V_1 = P_2V_2$$

$$850 \times 51 = 150 \times V_2$$

$43350 = 150 \times V_2$

$43350 / 150 = V_2$

$V_2 = 289$ cu in

To figure out how many times you can actuate a cylinder, you must know how much gas is needed to fill it when extended, the volume of the cylinder, and the V_2 value. The volume of an air cylinder is the area of the bore times the length of the stroke. Keep your units the same; if stroke length is in inches, make sure the area is in square inches. The formula is:

Volume = area × length

So, let's assume you have a 51-cu in tank supplying a 2-inch bore, 6-inch stroke cylinder at 150 psi. Given the formula above, you have 3.14 square inches for the area and 471 pounds for the force. The volume is 18.84 cu in. Double the volume for a retracting cylinder, and you get about 37.7 cu in. Divide total volume of the tank at 150 psi (V_2) by the volume of actuating the cylinder, and you get about 7 or 8 shots.

Several factors can make that 7 or 8 shots actually turn into about 15 shots. Temperature changes the pressure of a tank and system. Actuating the cylinder causes the CO_2 in the tank to "boil off" and create more gas. The boiling off draws heat away from the tank wall, which causes the wall to get cold. The water in the air surrounding the tank forms ice on the tank itself. The ice causes the temperature of the system to go down, which in turn causes the pressure of the system to go down. Less gas goes into the system because there is less heat to help the boil-off. Your cylinder is still completely open because the gas continues to expand no matter what, but the pressure of the gas is lower. That causes the force to go down. Thus, you lose power and gain shots.

The physical system itself may also cause power loss. Regulators allow only so much gas through at one time. Valves allow only so much gas through at one time. The inside diameter of the hoses and fittings carrying the gas constricts the amount of gas as well. One of the best things you can do to minimize system losses like this is to add a buffer tank.

Complete System

Figure 9.5 shows a working, if not competition legal, air system. In this example, we use CO_2 as the supply gas. The regulator is attached directly to the tank and changes the gas pressure to 150 psi. The hose from the regulator leads to the buffer tank through a T fitting. The buffer tank fills up in a very short time. The hose leading to the solenoid valve has a large inside diameter to maximize the flow rate. The solenoid part of the solenoid valve is connected to an electronic device that interprets the remote control signals sent out by the remote control receiver. When the signal to open the "extend" valve is detected, the solenoid does its job, releasing gas into the cylinder. The gas pushes the piston out until the end of its stroke is reached. The piston stays out as long as the solenoid is commanded to keep the valve open. Once the command to close the valve is detected, the solenoid releases and the valve shuts off. When the valve shuts off, the gas in the piston is released

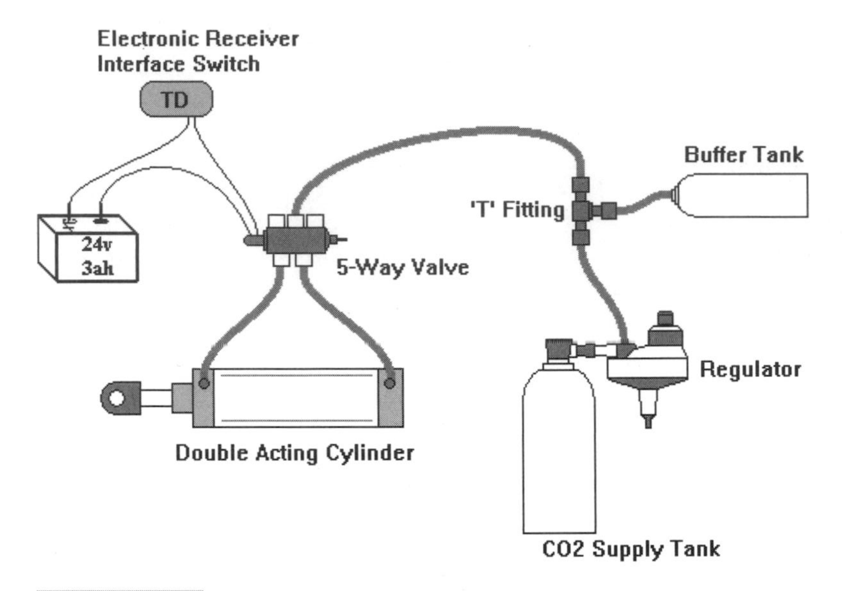

FIGURE 9.5 A working air system.

through the exhaust port on the valve. That would be the end of it if we were using a single acting cylinder. However, the cylinder in the figure is double acting and controlled by a five-port valve. So in this case, the command to close is actually a command to retract the cylinder. When the command to retract is detected, the valve controlling the second output port is opened and gas fills the opposite side of the cylinder. This causes the piston to travel back to its original position. When the command to extend is encountered again, the second port valve closes and vents to the exhaust port, and the first port valve opens once again, extending the cylinder piston.

Increasing the Force

Once you build a pneumatics system that runs on the standard 150 psi, you may want even more power. This is when things get a little hairy and certainly more dangerous, because you have to start using components made for *hydraulic* systems instead of pneumatic systems. You also have to buy the much more expensive solenoid valves that can handle the increased pressure. The walls of the hydraulic cylinders and hoses are all thicker than an air cylinder and hose walls. Hydraulic systems are built to handle higher pressures, but they can still self-destruct if you do not design the system correctly. In fact, lots of hydraulic cylinders have pneumatic ratings that are stamped right next to the fluid pressure ratings, because of the differences between gases and liquids.

A gas like CO_2 is compressible and acts like a spring in the cylinder. A liquid is not compressible and acts like a solid while in the cylinder. A good example of this is a hydraulic log splitter. The pump pushes oil slowly into the cylinder at a high pressure. As the splitter wedge contacts the log, the pressure is steady, and if there is enough, the log starts to split. Then the splitter wedge travels through the log at the same slow pace. If you were to pump air or CO_2 into the cylinder, the splitter wedge would slam into the log and stop. Pressure would build up, compressing the air like a spring until there is enough to split the log. At that time, the splitter wedge would release quickly slamming through the log and into the other end of the splitter. Hydraulic actuators are constructed so that they do not burst in a high pressure environment, but the ends of the cylinders just are not strong enough to handle the piston slamming into it at the end of the stroke. The secret to using hydraulic cylinders with pneumatic systems is to design so that the end of the stroke is never seen. Even that can be difficult, because you must design and build the frame of your bot to handle the violent end of stroke impacts.

Safety Precautions

A few things should be done to be safe when you are building and operating a pneumatic weapon system. The first is to never stand in front of your weapon or have any body part in reach of it. Never fill your system bottles and let them lie around. Fill them only when you are going to put them to use. Never use a higher pressure than the lowest rating of any component in your system. Never fill your own bottle unless you are qualified to do so. Finally, if you are new to pneumatics, do not try to design and build a system without the help of someone who has pneumatics design experience.

Summary

The most important fact that you should take away from this chapter is that pneumatic systems are dangerous and should be respected. Do not try to build or operate one if you are not qualified to do so. In Chapter 10, we briefly discuss the uses of gasoline engines and how to figure out how much fuel an engine will burn during a match.

10

GAS ENGINES

Several teams use gasoline engines instead of electric motors to power the drive system, weapon system, or both systems of their robots. A gas engine has a higher power to weight ratio than an electric motor and its batteries. However, there is always a trade-off: you not only get the extra power, but you also get a great deal more complexity and a great deal less reliability.

Complexity and Reliability

Gas engines are more complex and less reliable to work with than electric motors for a few good reasons. The more common engines, such as those found in chainsaws, leaf blowers, trimmers, and lawnmowers, usually only operate well in an upright position. To overcome this, you must design the bot to be a *self-righter* or use a *carburetor* that can run in all positions. Also, you need to include an electric start for these engines. There's nothing worse than getting hit one good time in the box and having the engine stall. You can't run into the ring and pull the cord again. On top of that, the engine may flood with gas while upside down and not be able to restart immediately. Another thing to worry about is the increased radio frequency interference (RFI) caused by the ignition system. Several people have overcome this drawback to some extent by using what is known as a *resistor spark plug*.

Gas engines are also more dangerous. To get their power, they harness controlled explosions. Gas is flammable.

Gas engines also have different power curves and spin faster than electric motors. You must adjust your drivetrain to handle each of these situations. While designing the gearing to handle that, be sure to include a way to get your robot to go in reverse, because gas engines don't reverse directions. I've seen some really clever ways to include gas engines in fighting robots, but it is my recommendation that you stick to electric motors for your first time out, even if you are a master mechanic of two- and four-stroke engines. If you do eventually decide to use a gas engine in your bot, be sure to read the Web sites of builders who have used them before. Do your homework on this subject and take your time to get it right.

Fuel Consumption

With all that said, I will pass on to you a way to figure out how much gas a given engine will burn during a match. You'll need this to figure out what size engine you can fit in your bot to stay within the fuel limitations. Steven Nelson, a bot builder with a lot of engine and fighting robot experience, figured this out. Here is what he said:

> I've been wondering how big of a gasoline engine I can run within the fuel limits.
>
> Here's what I found:
>
> Gasoline weighs between 5.8 and 6.5 pounds (lbs) per gallon.
>
> Gasoline engines burn between 0.4 and 0.6 lbs per horsepower (H.P.) per hour.
>
> 0.5 lbs per H.P. per hour is nominal (pph).
>
> There are 128 ounces (oz) in a gallon of gasoline.
>
> Using these three pieces of data, I've been testing some engine fuel requirements.
>
> Example:
>
> Engine H.P. equals 13.
>
> 13 H.P × .5 pph = 6.5 lbs of gasoline per hour (pph)
>
> 6.5 lbs of gasoline ÷ 6.5 pound per gallon = 1 gph
>
> 1 gph × 128 oz = 128 oz per hour
>
> 128 oz ÷ 60 = 2.133 oz per minute
>
> 2.133 oz × 3 minutes = 6.4 oz fuel consumption for a single match.
>
> 2.133 oz × 5 minutes = 10.66 oz fuel consumption for a rumble.
>
> Now, if I got the math right, 13 H.P. would be pushing it in the Heavyweight class but fine for the Super Heavyweight class. Of course, the pph estimates are for an engine producing full power for the total run time of the match. Usually you wouldn't run an engine fully loaded for an entire match. But these numbers are pretty good for estimating fuel consumption. Another thing to consider is that engines waste fuel until they reach their torque range and start to work

well. Try these numbers and see what size engine you think you can fit into the fuel limitations for your class. For a safety margin, I would use the 0.6 lbs per H.P. per hour. Your actual mileage may and probably will vary but not by much.*

Summary

I realize this was a very short chapter, but there is so much to talk about on the subject that I could not fit it all into the time and space constraints of this book. However, the major point of the chapter is that you must deal with the added complexity of a gas engine system and competition rules. As with pneumatic systems, I recommend that you do not attempt to build a robot that uses a combustion engine until you gain experience or enlist the help of someone knowledgeable in the subject.

In Chapter 11, we talk about common frame and armor materials. We discuss how to compare different materials, how to enhance the strength of lighter materials, and how to mount armor.

*Steven Nelson, Team SLAM; http://users.intercomm.com/stevenn/slamweb/slam1.htm

11

FRAME/ARMOR MATERIAL

Most robots to date have frames made from either steel or aluminum, although I have seen titanium, polycarbonate, and wooden frames. Usually, steel or aluminum frames are assembled by *welding*. Welded steel or aluminum is very strong, but you pay the price for either having it done or buying the equipment to do it yourself. *Extruded* aluminum forms can also be very strong and are easily assembled without welding. Damaged pieces can be replaced quickly instead of having to cut out, replace, and reweld a bent or cut piece of welded steel or aluminum. The drawbacks of the extrusions include the cost of materials and the required, specialized, and expensive brackets and mounting hardware. When it comes to making the choice of which material to use to build your frame, you must weigh the pros and cons of each material and make up your own mind.

The best thing you can do is look around for the best deal on the best material you can afford and work with. Many times, the difference between a hard material and a really hard material is a dent. The dents don't matter if your bot can still move and do damage. Many people use 6061 aluminum because it's readily available in surplus and relatively cheap. Still others use steel, stainless steel, titanium, or other materials. I personally have used many of the above materials in different capacities in all my bots. At least one bot frame has been made almost entirely of polycarbonate (bullet resistant plastic). (Lexan®, Tuffac®, and other names are different companies' brand names for polycarbonate.) These bots can use a material as flexible as polycarbonate because they have plenty of gussets. Put simply, gussets are extra materials placed in strategic places on the frame to add strength and rigidity. When building frames, it pays to observe the alphabet: Many letter shapes make excellent gussets for the main structural members. Letter shapes like A, V, X, and Y are

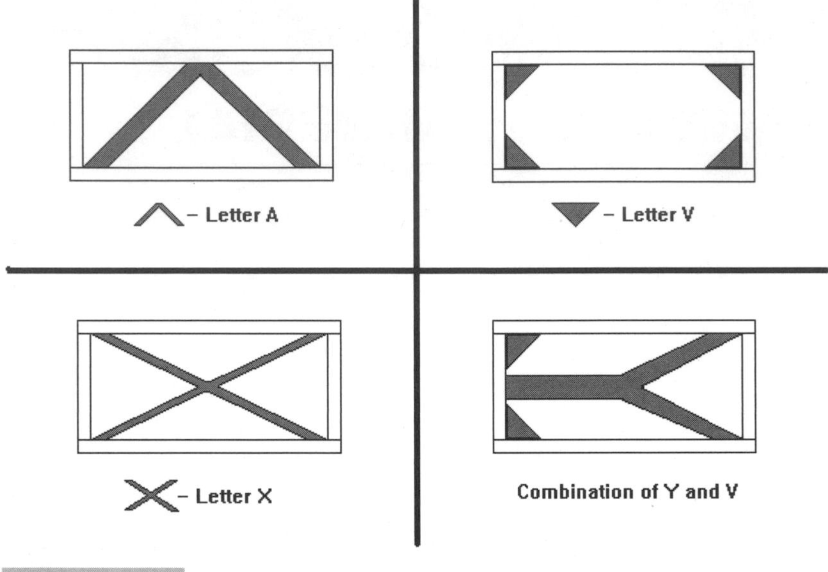

FIGURE 11.1 **Using letter shapes to strengthen a frame.**

very good strengthening agents and can be found extensively in buildings and bridges. Check out Figure 11.1 for some simple examples of gusset use.

Material Comparison

Material hardness ratings are relevant only when comparing one material with another. Any material can be bent or broken if it is hit hard enough. You don't necessarily want to use the hardest material that you can find, because hard materials are usually very brittle, not to mention expensive. Brittle materials break relatively easily when compared to softer materials. As with many things in bot building, material characteristics are a tradeoff. In this case, it is between weight, strength, hardness, elasticity, availability, and cost. With that said, Table 11.1 shows several materials and their ratings for comparison purposes.

In Table 11.1, you see the four keywords—tensile strength, density, hardness, and elasticity—along with their definitions, to use when comparing materials. For your use, there are charts of "Stock Aluminum Properties," "Properties of Metals," and "Weights of Materials" in Appendix C. I cannot include every known alloy and shape of every material. If the charts are missing some piece of information that you would find useful, I suggest taking a look at the Matweb.com Internet site.

Ultimate tensile strength (UTS), *hardness*, and *density* are probably the main items on which you should concentrate. Notice the UTS of 2024-T4 aluminum (68,200 psi) and of 1020 Steel (60,900 psi). Next check out their densities. The aluminum is 0.1 pounds per cubic inch (lb/in^3) and the steel is 0.284 lb/in^3. Thus, it takes about 7,300 more psi to break

TABLE 11.1 MATERIAL PROPERTIES

MATERIAL	TYPE	ULTIMATE TENSILE STRENGTH (PSI)	DENSITY (LB/IN^3)	HARDNESS (ROCKWELL B)	ELASTICITY (KSI)
Steel	1020	60,900	0.284	68	29,700
Stainless-Steel	304 Cold Rolled	73,200	0.289	70	28,500**
Titanium	Ti-6Al-4V Grade 5	150,000	0.16	—	16,300**
Aluminum	1100-0	13,100	0.0979	23 (Brinell)	10,000
	2024-T4	68,200	0.1	75	10,500
	6061-T6	45,000	0.0975	60	10,000
Polycarbonate	GE 101	10,000	0.0434	—	334 (flex)
Kevlar	DuPont 29	424,000	0.052	—	—

*Material ratings from Matweb (www.matweb.com).

**Value is the midpoint of the high and low ends of the range specified.

—Property not listed.

Ultimate Tensile Strength—Ultimate strength of a material subjected to tensile loading (until it breaks).

Density—The mass in pounds of 1 cubic inch of the material.

Hardness—The measure of a material's resistance to permanent deformation (permanently bent or dented).

Elasticity—Amount of stress you can put on a material without causing permanent deformation.

the aluminum, yet it is nearly three times lighter than the steel. Which material would you rather use on your robot? When comparing materials, you should get the manufacturer's data sheet for the specific alloy of material. Some material ratings are listed as values that can be obtained if you have the material heat treated. The true values are far lower without it. So, if you buy some S7 tool steel that hasn't been heat treated and you use it as-is, you just wasted the extra money you spent.

To me, material comparison is really not necessary. As I said before, you have to use what you are able to find, able to afford, and able to work. Finding different materials is not difficult using the Internet. The cost of materials is likely to set you back in the beginning, but the most important thing to consider is whether you have the means to actually put the material to work for you. If someone gives you a 4 × 4-foot sheet of 1/8-inch thick titanium, you must be able to cut, drill, and form it into something useful.

Sandwiches

What is the best way to make, build, and use armor? That's a difficult question to answer. It depends completely on your individual bot, your access to tools, and your money supply. In my opinion, the absolute best configuration is to sandwich exotic materials together to gain the benefits of each type. However, for a first-time bot builder, it may be easier to rely on readily available materials. I've used aluminum or steel as a body material and lined it with polycarbonate as extra protection for components that were mounted close to it.

Where possible, I mount the polycarbonate so that it's not in direct contact with the body material. My theory is that a saw blade, spike, or whatever is using up a lot of its energy to dent, cut, or break through the first layer of protection. The space between materials gives you a small amount of time to move the bot and put the opponent's weapon in a bind within the hole it just made in the first layer. This can drain the rest of the energy of the weapon. Very little movement is required to create the bind, as anyone who has broken a jigsaw blade, endmill, or drill bit will agree. The polycarbonate has its own excellent properties that resist damage. If your opponent's weapon gets through both layers without stopping, he has a better setup and you should design better next time. Many people get excellent results simply by combining two types of material. Some use epoxy to secure the materials together, some use bolts, some use both. See Figure 11.2 for examples of both methods.

Another sandwiching idea was handed to me by my friend–machinist. He mentioned that at some point in his career he had to cut a lot of lengths of steel round stock. The round stock came in bundles of 20 to 30 bars. His band saw was big enough to cut the entire bundle at one time. However, whenever you band round stock, you inevitably don't get all of the bars secured, or once you start cutting, a bar works loose. Once the saw gets to one of those bars, it stops cutting because the bar simply spins in place along with the saw blade movement. His solution was to spot weld all the bars together on the ends and then cut.

Sandwiched Steel and Polycarbonate Armor

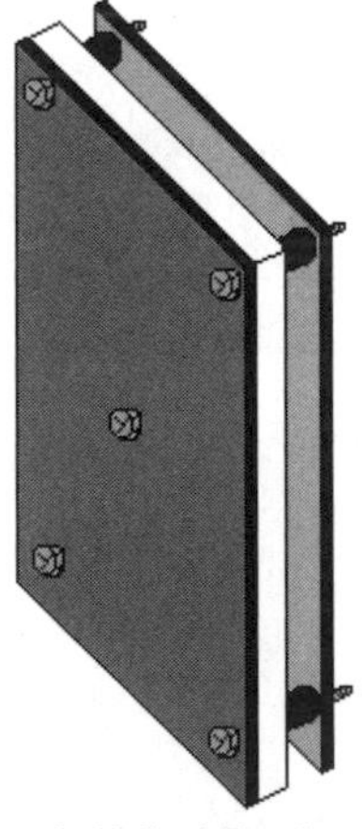

Sandwhiched Steel and Polycarbonate with a Shock Mounted Steel Layer

FIGURE 11.2 Sandwiched and shock-mounted armor materials.

The whole point of his armor idea was to use two flat pieces of material with round stock between them. Once a blade made it through the outer flat material, it would stop cutting because the inner round stock would start to spin. To be totally effective, you need several layers of round stock positioned so that no matter at what angle your bot met a saw, you would be able to stop the cut. This is impractical, because you can't fit that many sandwich slices in your armor. And because you need to have round stock big enough to not fit between the teeth of the saw blade, this idea is impractical, given the various types of saw blades used by builders and certain arena hazards. However, I would feel more comfortable just knowing that there was an extra protection layer or two between the opponent's weapons and my bot's electronics and batteries. Check out Figure 11.3 for a simple example of this roller-type armor.

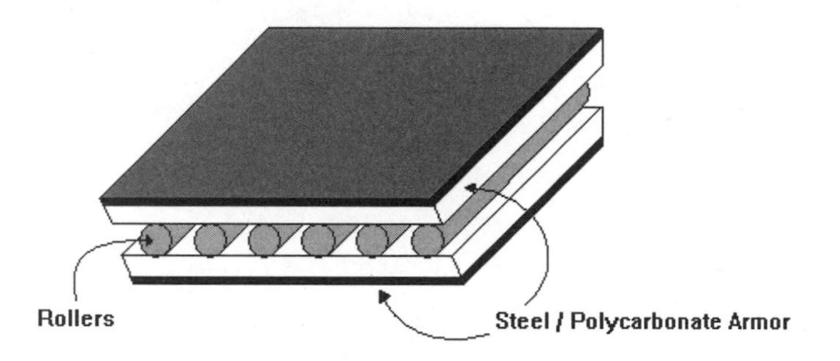

FIGURE 11.3 **Sandwiched round-stock armor.**

Shock Mounting

Yet another method of adding protection to your bot is *shock mounting*. Shock mounting attaches your armor so that when it gets hit, it moves and absorbs the hit and your frame does not. This is a round-about way of increasing the elasticity of whatever material you use for protection. For instance, one of my bots has an aluminum body, polycarbonate mounted on the inside an inch from the aluminum, and polycarbonate mounted an inch on the outside of the aluminum. The inner layer is mounted using 1-inch tall aluminum channels to space it out. The outer layer of polycarbonate is mounted using 1/4-inch tall rubber shock mounts that allows the polycarbonate to move whenever it is hit by anything. The rubber absorbs much of the shock that is created by an impact. So, for a hammer or spike weapon to get to the electronics of my bot, it has to penetrate two layers of 1/2-inch thick polycarbonate—one that's shock mounted—and finally one layer of 1/4-inch thick aluminum. It has to do this without getting in a bind within the 2 inches of air space between the materials.

Many kinds of rubber shock mounts are commercially available. Some are fairly cheap and some are expensive. I didn't want to spend money for something I felt I could easily make myself; a basic design for homemade mounts is shown in Figure 11.4.

I picked up some rubber bottle stoppers at the local gardening supply store. We drilled holes in the centers of the stoppers for the bolts. (If you do this, be careful when drilling rubber: it does not cut like metal. The rubber tends to stay attached and let the drill bit rip

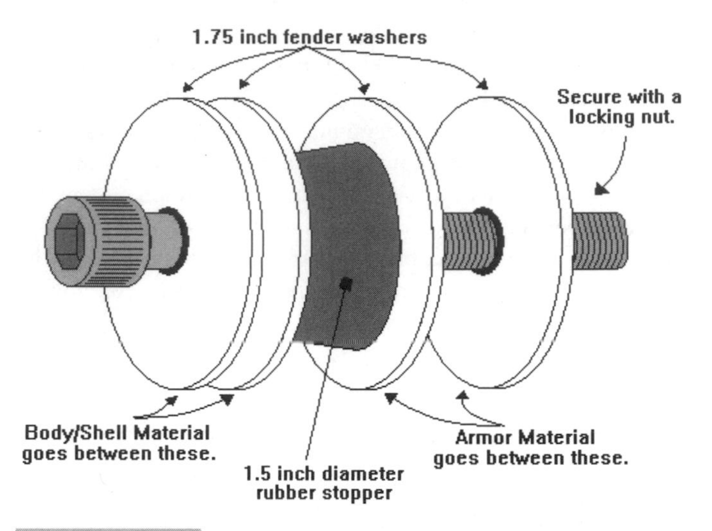

FIGURE 11.4 **Homemade shock mounting.**

through. Turn the drill speed up really high and plunge really slowly.) I also picked up some large diameter fender washer that nearly matched the diameter of the stoppers and the size of the bolts. First, mount the bolts in the main body material. Use fender washers on the inside and out so the forces are spread out. Then slide the rubber stopper on the bolt, followed by another fender washer. Put the polycarbonate—or whatever armor you are shock mounting—on the bolt, followed by another fender washer. Secure it with either a self-locking nut or a nut and thread locking solution. Tighten the nut down so that the rubber is compressed, but not so much that the rubber is ready to burst. Getting the tension right comes with experience and experiments. I usually tighten until the rubber is bulging slightly past the edge of the fender washer, depending on its original size.

Shock mounting is not only a good idea for your armor but also for your electric components. Speed controllers and batteries are susceptible to vibrations and can be completely destroyed by a good hit that jars your robot. It doesn't even have to penetrate. Shock mounting electronics can be really fancy. You can design a carriage and spring mounting system for all the electronic mechanisms and have them float in the middle of your bot. You can easily get similar, although not as good, protection from a low-tech computer mouse pad. Cut up a mouse pad and mount your electronics in small enclosures, with the pad as protection on the inside, outside, or both. Regardless of how you choose to protect them, it is not a good idea to mount critical components directly to the frame of your bot. Frames can bend and still function, but circuit boards are not made to bend. They will crack and your bot will die in its tracks.

Summary

In this chapter, we found that the type of materials we use may not matter as much as how we use them. Adding gussets to thinner, lighter materials improve their strength and reliability while saving weight that can be used somewhere else in your bot. Mounting armor so that it can move will improve its ability to protect the insides of your machine. Electronics should be mounted in the same manner.

Chapter 12 dives into the various weapon systems of combat robots. We discuss their weaknesses and strengths, and "hit" on a few ways to integrate different types of weapons on one robot.

12

WEAPON SYSTEMS

Many people who are new to this sport have dreams of mounting huge flame throwers and machine guns to their bots. Others think in simpler terms and create designs that implement drills, spears, and saw blades. Still others like to incorporate their weapon into the design of the body of their bot. There are infinite possibilities when it comes to dreams, but when it comes to actually building a bot, there are several tried and true weapon systems that fit within the rules. Many seemingly operable weapon systems are not actually doable; several other weapon systems just don't fit within the rules. Read the rules of each competition to determine which weapons you can use, and then read the Web sites of veteran builders to determine what is doable. If you have a weapon system in mind, try and find an example of it on Web sites before you attempt to build it. Many people think they have unique weapon ideas because they haven't seen them implemented on a TV show bot, but 99 percent of the time the idea is not unique or new. The other 1 percent is truly innovative. If you don't find your idea on a Web site, build a model to see if it will work. If it works, build the real thing and bring it to a competition. Research is the key to being able to implement the cool weapon systems that everyone likes to watch in action. If you find that someone has already built a weapon like the one you have in mind, build it anyway. Ask for their opinion on how it worked out and what troubles they had during the building process and while using it.

Weapons systems fall into a few categories, each with its own pros and cons.

Wedges/Rammers

The *wedge* is a favorite weapon of many new competitors because it is easy to implement. The traditional wedge bot is simply a drivetrain and a body shell that has one or more sides that slope down and come close to or meet the floor. The desire of most wedge bot builders is to tip their opponent off its wheels and slam it into the wall or into an arena hazard. Some wedges have enough power to flip an opponent by charging at high speed. In many bots, a wedge is the only weapon. Because the wedge is easy to incorporate and build, you will also see it in conjunction with powered weapons and used as a backup.

I include *rammer* bots along with the wedge bots because, like the wedge, the bot is the weapon. Many builders make their bot invertable, so that if it gets flipped over, it can still operate. Wedge and rammer bots are the easiest invertable designs to build, except that the one problem with invertable wedge bots is that their wedges are no longer effective if they get flipped. To compensate several competitors put two wedges on their bots: One wedge is effective right side up and the other is effective upside down, as shown in Figure 12.1.

Many people mount spikes or some sort of instrument on their bot to puncture or dent and use the tactic of getting up speed and slamming into a opponent. This often works better than anything else when it comes to disabling a opponent. On impact, electrical connectors can come apart, fasteners can vibrate loose, and structural members can lose integrity. The catch is that your rammer or wedge bot is taking the same beating as your opponent's. Most rammer and wedge bots have stronger drivetrains than bots with powered weapons, and therefore can withstand the same amount of abuse without breaking. Because the rammer or wedge can afford to use all of the weight limit to build a drivetrain and frame, they can use bigger, stronger motors and other components. Many people also shock mount the shell of these bots to minimize the amount of damage they take. The keys to an effective wedge or ramming bot are a strong frame, strong drivetrain, and plenty of driving practice.

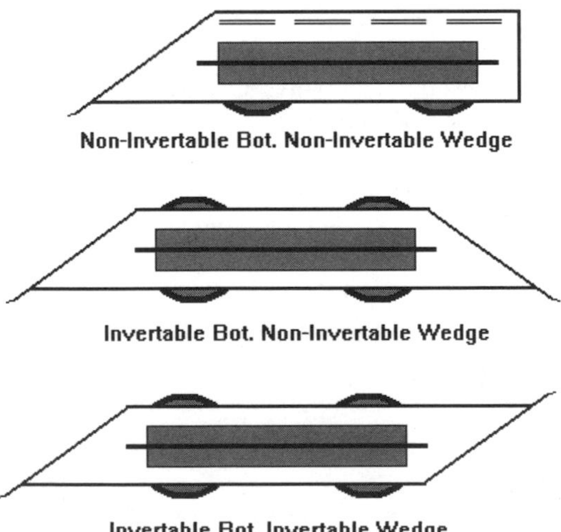

FIGURE 12.1 Invertable and noninvertable wedge designs.

Spinners

The *spinner* bot category is probably the most dangerous. Basically, these bots build up energy in some form of spinning flywheel and release most or all of it on impact with their opponent. Saw blades can be included in this group if they are meant to transfer energy and not simply cut the opponent. The difference between a blade meant for cutting and a blade meant for energy transfer is usually the size and rpm of the blade. Larger, heavier blades are best suited for beating on an opponent. Energy transfer blades can be spectacular, causing flying pieces of bots or entire flying bots. Spinner weapons take a lot of self-abuse, because of the amount of energy they transmit and the fact that it can be difficult to insulate the rest of the robot from the impact shock. Bots that have heavy spinning weapons sometimes have a difficult time steering while the weapon spins, because of the forces holding the spinning mass in its position. These driving problems are demonstrated in Figure 12.2. Within the spinner category are the horizontal and vertical subcategories. The horizontal spinner weapons don't experience the same amount of driving disruption because the forces are aligned with the drivetrain. They *do* experience driving problems when the spinning mass is not spinning around the bot's center of gravity. Vertical spinning weapons have the hardest time with steering.

Another disadvantage that the verticals endure and the horizontals do not is the fact that you must aim to hit your opponent. Many horizontal spinning bots are protected on all sides by their weapon, and some horizontal spinners use a shell that spins separately from the frame. Others spin the entire robot, although there is a driving disability endured by the "whole" spinners: it is very difficult, if not impossible, to manually control the direction of travel. Therefore, most whole spinners simply start spinning and wait for opponents to

FIGURE 12.2 Spinner driving problems.

come and destroy themselves. A few people have developed a computer-controlled method for altering the speed of one or more drive wheels at certain points of rotation to make the whole spinner creep along in one direction or the other while spinning, but it has not been perfected as of this writing. One disadvantage that all spinners must attempt to overcome is the fact that once they hit something, they must regain their spinner momentum. This can be a problem if their opponent is strong enough to take the initial hit and keep coming. The opponent can continue ramming, causing damage and gaining points, while the spinning mass is stationary or at a very low rpm.

Hammers, Spikes, and Clampers

I normally group all the powered hammers, powered spikes, and clampers within the same class. These weapons are generally a little more complicated to implement. They are also not usually as effective as spinners. Recoil times, weight, and speed all affect the destructiveness of these weapons. Fortunately, there are many ways to power them, thus giving builders the chance to experiment and find strong, effective, repeatable ways to deliver a punch. The common methods for powering hammers, spikes, and grippers include electric, pneumatic, and hydraulic actuators.

Each type has its pros and cons. *Electric actuators* can be small, high-powered, and easily controlled with simple electronics. They can also be expensive and have the same trade-off between torque and speed as electric motors and gearboxes, because they actually *are* electric motors and gearboxes. *Pneumatic actuators* can be lightweight, fast-acting, and fairly easy to control using special valves and simple electronics. However, you have to have a big one to put very much power into your punch. They are also complicated and can be extremely dangerous because of their speed and operating pressures. *Hydraulic actuators* are very strong and as easily controlled as pneumatic actuators. They are also very heavy, big, usually slow, and expensive. Correctly built, hydraulic-powered weapons can easily crush an opponent if you can catch them and get a good grip. Pneumatic and hydraulic actuators inherit the drawbacks of the systems that power them.

When dealing with pneumatic and hydraulic systems, you have to remember one thing: *flow rate is everything*. To increase flow rate, you want large inside diameter (ID) ports, hoses and components. Use components that are rated for what you want to do. Most common pneumatic components are rated for 150 psi maximum. When using pneumatics, you need to decide on which gas to use. Gases like nitrogen (N_2) and carbon dioxide (CO_2) are readily available and fairly inexpensive. (Never use a flammable gas.)

Pneumatic systems depend on a tank gas-filled. If the gas is CO_2, you usually depend on it to expand from liquid form to gaseous form, which requires heat. You can expect ice to form on the tanks after repeated weapons usage because the system draws the heat it needs out of the surrounding air. The expansion of the CO_2 or the pressure of the compressed N_2 powers your weapons.

When you use a hydraulic system, the gases are generally replaced by oil. The oil in these systems must be pumped to the valves and kept under constant pressure. To do this, either an electric motor or gas-powered engine is required. So, you have to not only deal with the construction and controls of the hydraulic end of the system, but also build and operate the electrical or gas-powered end, each of which has its own complexities and problems. The payoff for using such a complicated system can be great. However, please do the necessary research when you decide to build a pneumatic or hydraulic weapon. Do not just jump into the project blindly.

Flippers and Lifters

Lifters work by getting the other guy off his wheels. The more powerful wedges can be considered flippers if they are capable of turning their opponent over on its side or top. Most bots that are considered flipper bots use some type of leverage arm that is powered either electrically or pneumatically. So far, hydraulic power has proven too slow to implement a flipping mechanism but it may work simply as a lifting mechanism. Some flipper bots use fairly complicated designs that lift up and push out at the same time, and others use a forklift-type design. All flippers require the driving skill to position the flipping instrument and the ability to get under the opponent. The main concept to understand when designing a flipper bot is that of the lever.

A *lever* is basically an arm connected to a *fulcrum* or *pivot point*. Levers work in much the same way as pulleys—distance can be traded for perceived weight. Think of a teeter-totter where the plank is balanced in the middle on the support. If a 50-pound child sits on one end and a 150-pound adult sits on the other, the weight of the adult will lift the child into the air. Now move the teeter-totter's balance point, as shown in Figure 12.3, so that the distance from that point to the child's end of the plank is three times longer than the distance from the point to the adult's end of the plank. The child and adult balance out.

Besides being used as lifters, many other levers are used on fighting robots—you just have to know how to find them. For example, an overhead-swing hammer is simply a lever with a hammer-shaped weight attached to one end. The fulcrum or pivot point is where the hammer arm is attached to the robot. A motor and gears, which put out a specific amount of

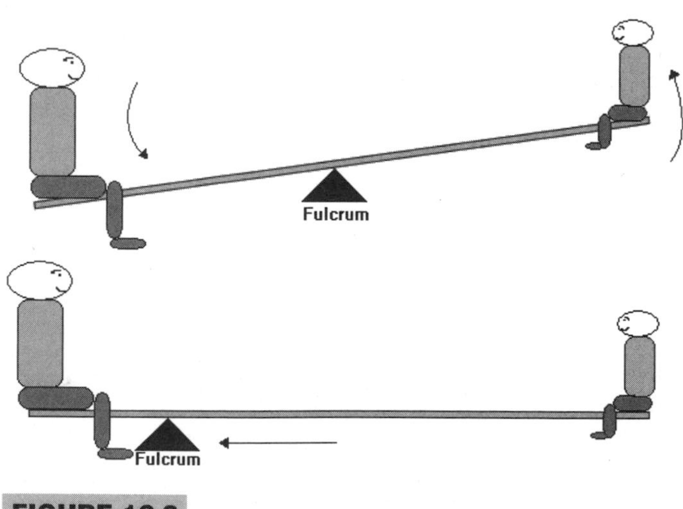

FIGURE 12.3 **Teeter-totter balancing.**

force, replace the other end of the teeter-totter plank. Usually the amount of force the motor and gears put out exceeds the weight of the hammer by a good bit to improve swing speed.

Miscellaneous Weapon Types

Some competitors want to include smaller saw blades, milling machine cutters, and drill bits into their weapon designs. Although it is true that some of these weapons can cause a good bit of sparks and gouges when in contact with the opponent, they really don't do the damage that we all want to see and that is required to effect a knockout win, because all these weapons require constant contact in one spot on the opponent to do real damage.

One type of saw blade, however, does a good bit of damage: emergency rescue saw blades. These have big teeth and are made to cut through just about anything. One competition that I know of uses them as an arena hazard. However, these emergency blades are expensive and they must be powered by a strong motor or gas engine to be effective.

Another favorite of the new or hopeful builder is to attempt to use an electromagnet to engage an opponent and hold it still long enough to use a saw or drill. Once you go to a competition and see the bots in action (or just read the Web sites), you realize that most bots aren't made completely of steel, and on several materials, magnets have no effect. Even if the opponent's shell is made completely of steel, it is likely that the shell will have odd angles that make it difficult for the magnet to grasp. Even if the opponent's shell is completely made of steel and formed with flat sides and tops, you will need a very powerful magnet to hold it—so powerful that I would wager you could not put enough batteries in the bot to power the drive, power the weapon, have a truly effective magnet system, and still stay within the weight limits of any weight class. Magnets are best left inside motors.

Summary

Weapon systems are a lot like a game of rock-paper-scissors (or hammer-wedge-spinner according to some builders). Many builders try to incorporate two or more weapon systems into one bot, but the major problem with this approach is that you have to sacrifice some strength to get more than one system into the bot.

In Chapter 13, we dive into some more theory about robot design. We answer seven questions that are the keys to good design, and we also cover the top four categories of failures that robots experience during a match.

13

DESIGN TIME

No single, perfect robot design exists. In the years that I have been involved in this sport, I have only witnessed two real innovations in robot design: the vertical spinner weapon and the *clamper* robot. That may change now that so many more people are getting involved, but for the most part, people take existing designs and attempt to modify them so that they do more damage or take less damage from an opponent. Many competitors mix the good points of one design with the good points of another design to create a better bot. It is my opinion that new builders should dump their fancy, killer designs in exchange for something simple. Build something that works and compete with it so that you gain experience. When I was new to the sport, I did not think the veterans had that big of an advantage. I was wrong. Watch what they do and how they do it. You will learn a lot. After that, go build your dream machine.

Seven Important Questions

There are seven questions to ask yourself when you set out to design a fighting bot:

1. What are the rules of the competition I will be attending?
2. What materials can I use to build my menacing machine?
3. What strategy will I use?
4. What weapons will it have?

5. How will I control it?
6. Can I pay for what I am designing?
7. Do I have time to build it?

You will be scratching your head and asking a lot more questions during the building process, but for now we will concentrate on designing.

WHAT ARE THE RULES?

Download the rules for the intended competition from the Internet. I have not heard of a fighting robot competition that did not have the rules available for download. If you do not have access to the Internet, get it. The Internet is probably the most valuable tool you can use when building robots. Public libraries, schools, and Internet café houses are places you can go to get online if you do not have a computer at home. The point is, know the rules of the game so that you can compete when you get there. Knowing the rules also helps you to design a bot that fits a certain strategy. Rules dictate how long you have to spend repairing a broken robot and charging batteries. They also dictate how long the match lasts. You should keep these time limits in mind during the design phase.

WHAT MATERIALS CAN I USE?

Before answering this question, you must decide the weight class in which you will build. This decision affects which materials, motors, wheels, and batteries you use. After deciding the weight class, the materials question is answered in three parts. Part one says that the rules of the competition usually list the types of materials that are not allowed on the robot. Part two says that you cannot use a material if you do not have access to it or a way to work it. You may have access to free titanium sheets and rods, but if you do not own or have access to equipment that can cut and form it, then the material is useless. So, unless you have access to a machine shop that can handle titanium, stick with aluminum, steel, and polycarbonate at first. Part three says that you must use what fits inside the weight limit. If you have a piece of 1-inch thick steel plate that is 2 feet square, you will not be able to use it as armor on an effective fighting robot. Depending on the grade of steel, this piece will weight more than 160 pounds by itself. Even if you are building a super heavy robot, up to about 340 pounds, you don't want to use up almost half of your weight allowance for one small piece of armor.

The bottom line is that you do not want any unnecessary materials on your bot. Extra weight does nothing but slow the bot down. Large electronics, long runs of heavy wire, oversized parts, and unnecessarily thick armor add up very quickly, and you will find yourself struggling to lose weight. When this happens—and it will—sit back and take a long look at the robot. Look for places that have really thick supports where you do not need them. Remember that some shapes are stronger and lighter than solid pieces and might fit in the same place. I once cut 66 pounds off a robot to meet the weight limit, after discovering a lot of unused metal that I could take out. One trick was to bore out all the shafts that were 1 inch or larger in diameter. Many builders drill holes in their frame material and supports to drop the weight. When you have finished losing weight, sit back again and look for more heavy spots. You will find some.

Stay simple when selecting the materials you will use in your bot. It will pay off later when you are running against a registration deadline.

WHAT STRATEGY WILL I USE?

Strategy is an important aspect of bot building. Different types of bots utilize different strategies. For instance, a bot with a vertical spinner weapon must be quick as well as strongly built so that it can keep a weapon aimed at the opponent. A bot that spins its entire body horizontally does not need to be quick, but it needs to be very strong. Ramming bots must be fast and strong. Lifting bots must be precise. This is called *offensive design*. In the *defensive design* category, you must design to be effective against a large number of weapons and strategies. Can your armor withstand a hit from a spinner weapon? Can your bot drive while upside down or turn itself over if flipped? Can your opponent get under you? All these questions get answered as you decide on your strategy.

Depending on the type of competition you are going to, you may need different design strategies. If a competition focuses mainly on controllability and power, there is a completely different design strategy for it as opposed to the competition that focuses solely on fighting. I am not going to give you a strategy to follow to beat each type of opponent you might face or to win each type of competition you enter, because I don't know them all. Your job is to come up with a strategy and see if it works. That is part of the fun of the sport.

WHAT WEAPONS WILL IT HAVE?

This question is answered while answering the strategy question. If you are going to put a weapon on a bot, make sure it can do some damage.

Some criteria:

- You should not want to go near the weapon when it is armed or running.
- It should be so fierce that you question the intelligence of turning it on while the bot is in close proximity of people or animals.
- It should really scare you.

If a weapon is not scary to you, it will do nothing to your opponent's bot. If it will do no harm to your opponent's bot, there is no point in having it. That said, be careful when building these things. Build them strong enough to handle their own forces. If it looks flimsy or weak, it is. If you can bend it by hand, it is too weak.

A limited number of weapon types is allowed in any of the competition rules. The most effective ones are the impact and lifting weapons. Cutting weapons usually do not cause much damage, but a large saw blade can be used as an impact weapon instead of a cutting weapon. Impact weapons include bots that ram their opponents, and some of those have spikes for impaling. Spinning blades, disks, and entire bots are also impact weapons. Bots with swinging hammers, chisels, or spikes are also in the impact weapon category. They all have one thing in common: they all try to store as much energy as possible and release it into the opponent as quickly as possible. Impact weapons have another thing in common: they must be robust. An impact weapon can cause as much damage to itself as it does to the opponent when striking.

The lifting weapon category obviously includes any bot that has a lifter arm meant to get under or grab an opponent and bring it completely or partially off its wheels. However, a simple wedge bot is also in the lifting weapon category. By its very nature, it attempts to get under its opponent and take it off the ground.

No matter which weapon you chose, there is a bot out there that will be destroyed by it. There is also a bot out there on which it will have very little effect. Some of the best designs that I have seen have utilized two or more types of weaponry. A bot with a vertical or horizontal spinner blade can also have a wedge incorporated into the body. A bot whose main offensive objective is to push the opponent around with a bulldozer blade can also have a spinner weapon. The main things to remember are to keep it simple and build it to last.

HOW WILL I CONTROL IT?

Refer to Chapter 2 for descriptions of the different types of control mechanisms used to run a fighting robot. The main thing is to get something that you are comfortable using. If you have been an RC car hobbyist for a while and really like the trigger or steering wheel type of transmitter, go with that. If you fly model planes or helicopters and are comfortable with the standard two-stick remote transmitter, go with that. There is even a 900-MHz system that uses a standard PC joystick like that used by computer gamers.

If you have more than one team member, decide early who the driver will be and make him/her practice—a lot. I do not recommend that he/she be the only person getting driving practice, in case he/she cannot make it to a competition down the line and someone else has to drive.

If your weapon is complicated, which it should not be if you are new to the sport, or if it is just plain hard to drive the bot and fire the weapon accurately at the same time, you need some way of controlling the system using two people. Many higher-end airplane and helicopter radios have a *trainer jack* on the back, so that the person who knows what he is doing can control the transmitter of the person who is just learning. Many builders get two transmitters and connect them via the trainer jack so that one can drive and one can activate the weapon. Other builders use two RX/TX setups in the same bot. One controls the drive and the other controls the weapon.

CAN I PAY FOR IT?

This is an important question, and many factors and expenses must be considered when answering it. Many first-time and young builders stick to the lighter weight classes to save some money. A 300+ pound robot will cost you a lot to build: Motors are expensive. Speed controllers are expensive. Remote control units are expensive. You also have to think about how you will transport a 300+ pound robot to the competition. Sponsorships can help, but do not count on a company to just hand money over to you if you have never done this before. (Check out the "Getting Parts" section later in the book to help you get a sponsorship.) The best thing you can do is design your robot and look for all the parts you will need to build it. Add up all the prices and see if the total fits in your budget. If it doesn't, design something simpler and smaller. Leave the bigger design for later.

DO I HAVE TIME?

If you work 60 hours a week and have a family who needs your attention, you probably do not have enough time to build an effective fighting robot. Most competitions don't occur on a regular basis. The event sponsors usually announce one and then wait a while to announce the next one. The longest lead time I have seen is 4 months. However, that amount of time is rare. Some competitions only give you the specifications for the event 8 weeks ahead of time. They do this on purpose, as part of the challenge. As you become a veteran robot builder, it takes a lot less time to put something together that will last. My first competing robot took more than 6 months to build, and it still suffered devastating problems at the competition. One of my latest bots took only 3 weeks to build and did extremely well at competition time. Both bots weighed more than 200 pounds. I know one veteran builder who put together a really effective bot in just 1 week. If you are new, estimate how long it will take to build your bot and then double that. Don't forget to include shipping time for parts.

Overall Design

When you design your bot, do not try and come up with one that will kill everything in the ring. You cannot do it. I know of only one undefeated bot that is still active as I write. I hate to say it, because the bot is one of my favorites, but the laws of probability will bear down on that bot before too long. Be realistic and create a design that you think will be effective against as many types of bot as possible. If you cannot do that, create a design that has several good points and as few weaknesses as possible. Then go have some fun. Remember the basics of armor and protect all your sides, including the top and bottom of the bot. If you can design modular weapons—weapons that can be changed out according to what opponent they would hurt most—that is a plus. However, this approach can get really complicated and is probably too ambitious for an inexperienced builder. Remember to mount your electronics so that vibration from opponent contact does not affect them.

Most Common Failures

The most common failure points of combat robots are numerous yet they fall into about four categories. The "parts that are not used correctly" category includes speed controllers, batteries, motors, and gears that are expected to do more than their designated ratings allow. If a motor is too weak, it is too weak. Do the math, and do not expect it to pull the weight if the numbers do not match.

The "parts that are not mounted correctly" category includes everything that you bolt, strap, tie, or use bubble gum to stick to the walls of the robot. Bolts should use some type of thread locking solution and should be checked for tightness after each match. Major components such as batteries should have no possible way to come loose and flop around. Sprockets and gears should be lined up properly. Wires should be crimped correctly.

The "parts that are not reliable" category includes gas engines, home-built electronic components, and improper remote control antennas. If you know that your tires can be punctured and cause a problem, use solid or foam-filled tires.

The last category, "bad design," can be helped only by experience. If you need ground clearance to compete, make sure you have enough. Things like that will make you smack your forehead. Pay attention to all four categories, and you will have a much better chance at winning something.

Finally, remember that no design stays the same from drawing board to rolling on the floor. You will change some aspect of your design almost every time you work on the robot. You will make a part, only to notice that the part next to it needs to be a different shape. You will make a part that looks good on paper but just does not work right in real life. Sometimes you won't be able to get a store-bought part that you had planned on getting. Even after you build and compete the robot, you will notice things you want to change to make it stronger or more dangerous to an opponent. None of this should discourage you. The design and build process is a major part of the fun.

Summary

When designing your robot, answer the seven questions as honestly as possible. There is a very large difference between the robot of your dreams and the robot you can actually build. I don't say this to be condescending; building that dream robot is something that can take many years of study and hard work. If you pay attention to the common failure points, create a robust design, and get some driving practice, you will have fun at competitions—you may even win some fights.

In Chapter 14, I cover several ways of making sure you have a good physical design. Physical design is different than strategic design. When I speak of physical design, I mean parts placement. Anyone can draw a robot on a piece of paper and call it a design, but not everyone can design a bot from inside out. I cover a few ways to help those of us who find it difficult to place parts on paper.

14

MODELING

Lots of people, including seasoned vets, start out with simple drawings and design from the ground up. Ideas can come at any time. I've got notes and simple sketches on the backs of envelopes, match books, and napkins. Drawing something out is always better than letting it stay in your head. Once it's on paper, you can see it more easily. You can see the design flaws a lot quicker. When you have your design drawn out the way you think it will work, the next step is to build some type of model. Remember that a model does not necessarily have to *physically* exist.

Low Tech Models

Some people like to build small models from Erector sets or Legos. Lego and Erector set type modeling is good for making sure your design will actually work as planned. With the advanced kits out today, you can build working models that closely resemble the real thing. These will show you the problems inherent in the design, so that you can change it before spending your hard-earned money building something that won't work as planned. Figure 14.1 shows a small, Lego-style bot.

The next step is to see if you can fit all the pieces of the bot in the shape you have designed. This isn't too hard and can be accomplished quickly in a couple of ways. You can build cardboard models of all the major parts, like batteries, motors, speed controls, and structural elements. Put all these together with tape and see if it looks like it's supposed to.

FIGURE 14.1 Cyber-Scorpion. (Courtesy of Tony Hall at www.teamradicus.com)

Figure 14.2 shows a small bit of "cardboard-aided design" that I did for my 340-pound bot, RipOff.

Once you have built a few bots, you have a better feel for this type of thing. You also have more stuff lying around. When that happens, draw your outline on the floor with some

FIGURE 14.2 Cardboard-aided design.

FIGURE 14.3 **Chalk-aided design.**

chalk and place actual parts inside it to make sure of the fit. Figure 14.3 shows a sample of "chalk-aided design" I did for one of our 220-pound robots called HandsOFF!

High Tech Models

Some people skip all the physical modeling and go straight to the computer. They use specialized software to build models of the individual parts in their robot and place them in three-dimensional space. Figure 14.4 shows a Rhino3D screenshot of a small robot I never got around to building. Once done, you can move around a fully built robot on the screen and make sure everything is going to fit correctly. Modeling in the computer is necessary when you design parts that require a computer-controlled mill or lathe to build. Unless using a *CNC machine* is your profession, you will probably want to get a professional to do the work for you, and he may even need to model the part for you. Pay attention to what professionals say and do; they have years of experience that can benefit you in current and future designs.

Rhinoceros (Rhino3D for short) is a Non-Uniform Rational B-Splines (NURBS) modeling software for the Windows Operating system. NURBS uses mathematical representations that can define shapes ranging from a two-dimensional line to complex three-dimensional surfaces or solids.

Using Rhino 3D, you can create and edit on screen any shape you have the patience to design. You can also include dimensions or have the computer measure for you. It has the capability to render your designs in 3-D with colors, textures, lights, and shadows. It sup-

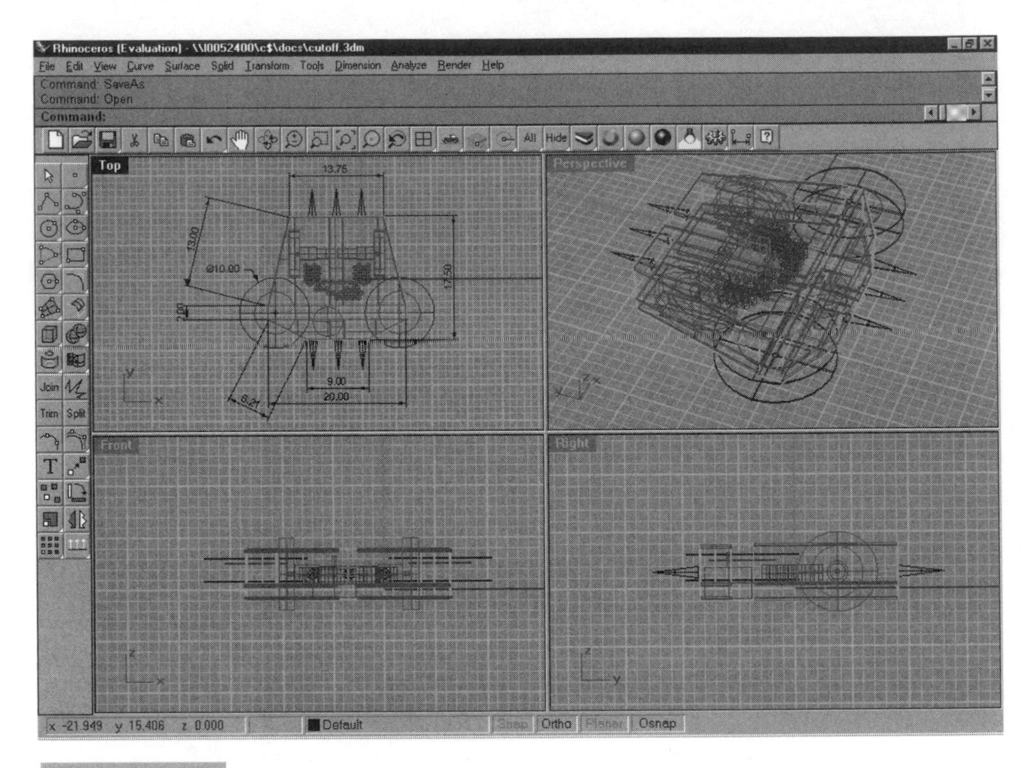

FIGURE 14.4 Rhino 3D screenshot.

ports many different file formats that you might find useful. It's designed to run only on Windows 95/98/NT/2K/XP. I know they do not plan to port to UNIX or MacOS. Rhino requires a Pentium or higher processor and more than 32 megabytes of RAM are recommended. About 20 megabytes of disk space are required for the software, not including design files, and no special graphics cards are required. Several tutorial files can be found on the Rhino Web site. Going through these files takes a couple hours, but in the end it is worth it. Rhino3D is the best and easiest to use design software that I have come across. Some high-end features are missing, but the amateur bot builder usually doesn't need them.

Rhino3D isn't the least expensive design software I've seen: it runs close to $900 for the average person. Students get a deal at about $300, and schools get a price break for labs as well. I have included an evaluation version of Rhino3D on the CD-ROM in the back of the book. The evaluation version isn't crippled in any way when it comes to functionality. You are simply limited to the number of times you can save a file before you have to purchase the software.

Computer-aided design (CAD) software like AutoCAD and ProE is a very small step up from Rhino for the hobbyist. These packages have very few extra features that the amateur will want to use extensively. Current pricing for AutoCAD2000i is just over $3,000. ProE runs about $21,000. Both packages are excellent for designing robots, and some engineers may feel that they are necessary but I do not. You can build a fun and competitive bot without drawing the first line on a computer.

Several very inexpensive software packages on the market allow you to do simple design work at a really low price. When I first started out, I used them for general layout and overall design. Figure 14.5 shows a screen shot of Microsoft's Paint program which is included with every version of Windows that I know about. You can draw many designs with Paint. However, if you want to know how wide an opening is on that design, you will have to build it and see. When using these programs, it's also sometimes more difficult to model 3D images. These programs sometimes lack several features that make designing your robot easier, but they are a good way to visualize it without building it first. If you want a program to get your idea visualized on the computer screen and not necessarily engineer the entire thing to specifications, you can use MilkSHape 3D, a shareware program that was written to build polygon models for 3D games. It has all the basic operations like select, move, scale, rotate, and extrude, just to mention a few. Primitives like boxes, cylinders, and spheres are also available. It has limited animation abilities. To find it, try a software search on the Internet.

Summary

Modeling your bot can be a great time saver. It can also be a great waste of time. Although I have committed most of my robot designs to the computer, they rarely end up exactly like

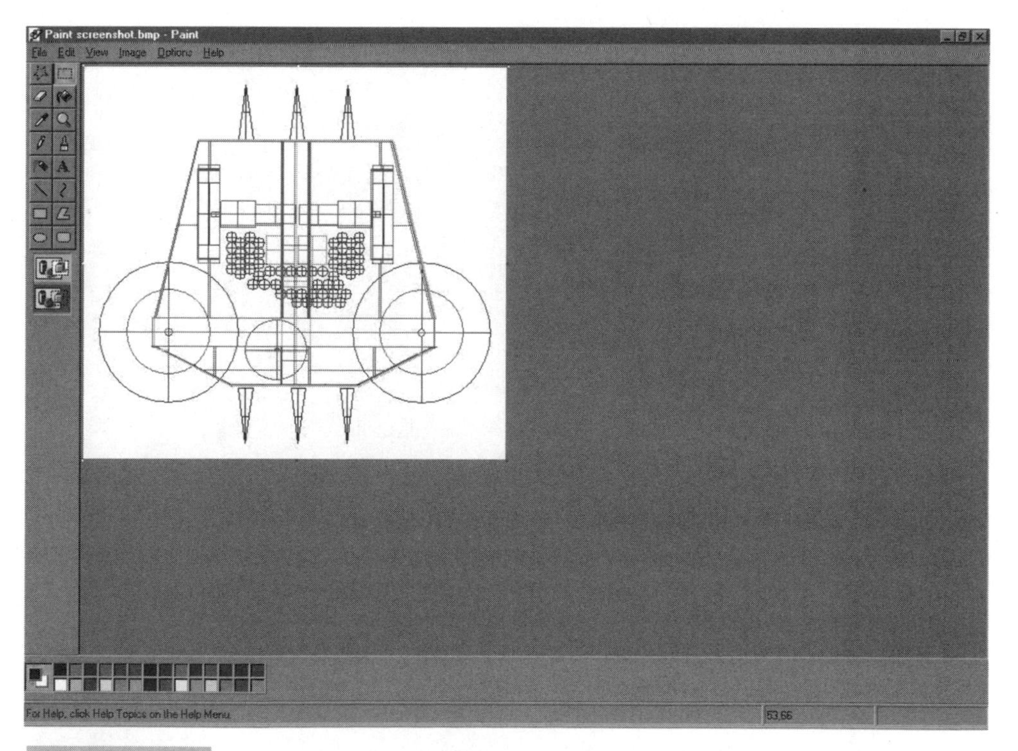

FIGURE 14.5 **MS Paint screenshot.**

the model, because I do not have computer-controlled equipment nor do I have the patience it takes to create something with such precision by hand. Instead, I put the main objectives into the model and set out to create something that closely resembles it. One of these years, I plan to build at least one robot completely from computer designs. In the mean time, I'm having too much fun actually getting something built.

In Chapter 15, I show you several tools and describe their uses. It is not necessary to own every single tool in my list. In fact, I don't own every tool listed—the mill and lathe just won't fit in the shop. Tools are meant to make a job easier. Not having a specific tool does not mean the job cannot be done. It just means that the job will take longer.

15

TOOLS

This chapter lists those tools that I've found useful in my endeavors and gives a short description of how to use them. I'm in no way saying to go out and buy all of them. After all, no one ever has all the tools he will ever need—there is always something that you could have done more quickly or easily if you had a different tool. What I will say is that you should buy the best quality that you can afford. Many differences exist between cheap tools and good tools. Using good tools is also generally less stressful, because cheap tools usually do not last. They are substandard to get the price down. So buy the best you can and keep them forever, if possible. Also, buy tools that have a lifetime warranty. These companies will replace any tool you break with no questions asked. Twenty years from now, you may break a tool and be glad you bought the good stuff when they give you a brand new one and apologize for the inconvenience.

Work Bench

A strong, steady work bench, like the one shown in Figure 15.1, is a must when building fighting robots. It is especially important when you start building the big ones. In our shop, we've gone from a plastic, poolside table, to one of those 8-foot long, folding-leg tables, and finally to a metal shop table with a 2-inch thick wooden top. The plastic table was a bad idea, even though it held up under the 200 pounds of one of our heavyweight bots. The folding-leg tables are still in use but mainly hold parts, batteries, and bolts. We had been

FIGURE 15.1 **Work bench/table.**

using a table saw as an assembly and welding table, which was OK, but every time we wanted to cut some material we had to move everything. It was also small. We picked up the true shop table with the wooden top at an auction. This is table is perfect. It also has drawers where we store our frequently used tools.

Bench Vise

A vise, such as the one shown in Figure 15.2, is very valuable in your robot building adventures. You use it to hold parts for cutting. You use it to hold entire assemblies for testing. Among other things, I've used it to press bearings into material. When it comes to a vise, bigger is better. If you buy a small one now, you will eventually buy a big one later. Spend the money on something with at least a 5-inch jaw. Because there are teeth on a vise, get some copper faceplate covers. The covers are softer than the faceplate and the teeth will deform them instead of your part. They also protect the faceplates when you are hammering on them. Mount the vise on your work bench, out of the way, but in a position that is centrally located.

Hand Drill, Drill Bits, Taps and Dies, and Reamers

You are going to drill a lot of holes while building bots. Figure 15.3 shows some assorted electric hand drills. The only recommendation I'll give is for you to get a hand drill with

FIGURE 15.2 Bench vise.

multiple speeds and reversibility, which should not be a problem since most drills today come with those features. Also remember that there are times when a cordless drill just will not do the job.

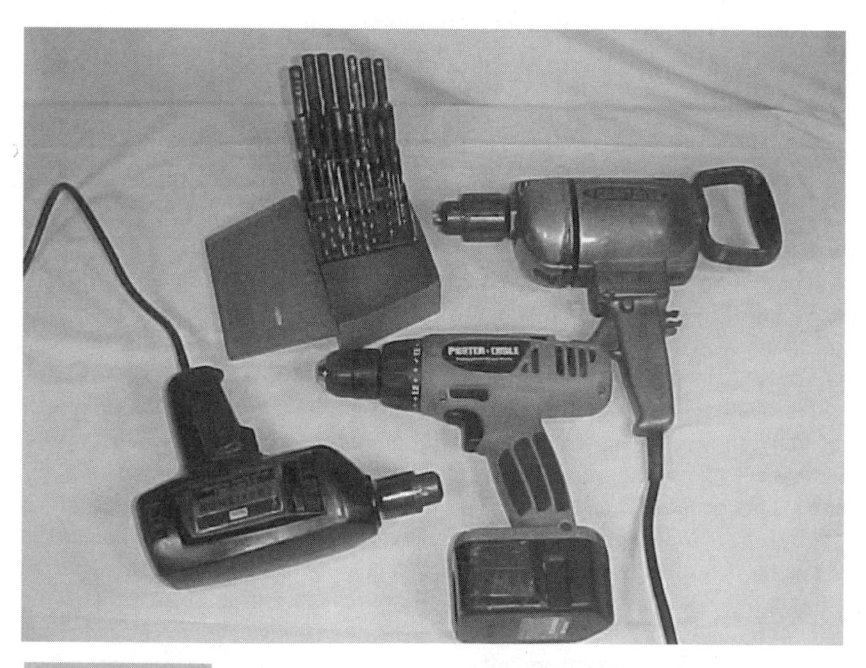

FIGURE 15.3 Assorted electric hand drills.

PILOT HOLES

When drilling holes in steel that are larger than a 1/4 inch, drill a small pilot hole first. I use a small bit that is as big as the solid center portion of the large bit (also called the "web"). The reason for pilot holes is simple: The point on a large drill bit does not like to cut, so you remove that material first. When drilling large holes you are removing a lot of material. When you remove a lot of material, the bit may grab and make an oblong hole. A pilot hole removes a bit of material and makes it easier for the larger bit to cut. Remember to use some type of lubricant to keep the bit and material cool and pay attention to the recommended drilling spindle speeds. The table of "Spindle Speeds for Drilling" in Appendix C tells you how fast the spindle should be turning for different types and thickness of materials. The "Tapping Lubricants Chart" in Appendix C can be used to determine the best drilling lubricant to use for different materials.

DRILL SIZES AND THREADS

You may want to bolt something to a piece of material without using a nut. In that case, you will have to tap and cut threads into a hole in that material. Industry standard sized taps require a hole of a specific size. That's when you need a drill and tap chart, like the table of "Drill Sizes and Threads" in Appendix C. It tells you the correct size of hole needed for tap sizes up to 1 1/8 to 7. You should also use tapping fluid, so remember to check the "Tapping Lubricants" chart in Appendix C.

THREAD CUTTING

To tap a hole, find the required hole and drill size on the table of drill sizes and drill it. Secure the part in a vise. Drop a couple drops of tapping fluid on the tap and in the hole. Start turning the tap in the hole slowly and absolutely straight in relation to the hole, as shown in Figure 15.4. After a couple of turns, make sure it is going in straight. If not, start over. Once you make a couple of turns and feel the tap cutting the material, reverse the turning direction of the tap (back off) by a half turn or more. When you hear some snapping or clicking noises, these are the chips of metal breaking free. Continue turning one to two turns, back off, and add fluid until the hole is threaded. Different materials and different sized taps require more or less backing off so that the chips can be cleared. Small taps require a lot of backing off. If it gets difficult to continue the cutting direction, start backing off more often and keep using the tapping fluid. Sometimes it will be necessary to back the tap completely out of the material and blow the chips off the tap and out of the hole.

Different materials hold threads differently. Steel can hold a thread under extreme forces. Aluminum and polycarbonate don't hold threads as strongly. When threading softer materials such as aluminum or polycarbonate, increase the number of threads in the material. This does not mean you should use a fine-thread bolt instead of a coarse-thread bolt; instead, a better plan is to use two or more bolts where you planned on using only one. If the situation permits, you could also use a longer bolt and put more threads into the same hole by tapping deeper into the material. Yet another solution presents itself with *thread inserts*, which have different forms. Some look like springs and are inserted into an oversized, threaded hole. These inserts give the bolt a steel thread to hang onto while gripping the softer material on the outside. Some inserts take the form of weld nuts or PEM nuts,

FIGURE 15.4 Tapping a piece of material.

which are described in the "Fasteners" section of the book—the basic idea is that the nut completely replaces the threads in soft or thin material.

Occasionally, you want threads on a small piece of round material. If you don't have a lathe and the part is small enough, you can use a die, as shown in Figure 15.5. Dies are much like a tap that's been turned inside out or like a doughnut where the inside of the hole

FIGURE 15.5 Using a die to cut threads.

is threaded. The same procedures apply when cutting threads on the outside of something as when cutting threads on the inside.

Good quality drill bits are a fairly expensive but an advisable investment in tools. Most drill bits can be sharpened at home using a bench grinder. To sharpen a drill bit, make sure the grinding wheel face is straight. Notice the angle of a sharp or new bit's point (about 60 degrees). Hold the dull bit at the same angle, with the cutting edge of the angle pointing up and parallel to the grinding wheel face, as shown in Figure 15.6. Hold the bit with both hands, one on the shank and one on the spiral. Gently bring the cutting edge of the bit against the grinding wheel face and slowly lower the shank of the bit while holding the front of the bit steady. Pull the bit away from the grinding wheel and rotate it a half turn. Repeat grinding until the bit is sharp and the angle is correct. Check it with a drill point gauge if you have one.

REAMERS

A reamer is a tool used to make really accurate holes. It looks like a drill bit that has straight cutting edges down the length of the bit instead of spirals. It does not have a pointed tip. Reamers are used in lathes, not hand drills. (I suppose they could be used in a drill press though I've never seen it done because you generally don't need such an accurate hole size when drilling a bolt hole.) Instead, a reamer is used to ensure the accuracy of a hole so that a small bearing, bushing, or sleeve fits properly inside a part. Ideally, you should drill the hole with a drill bit in a lathe to about 1/32 inch undersize and then finish with a reamer. Use oversized and undersized reamers for slip or press fits.

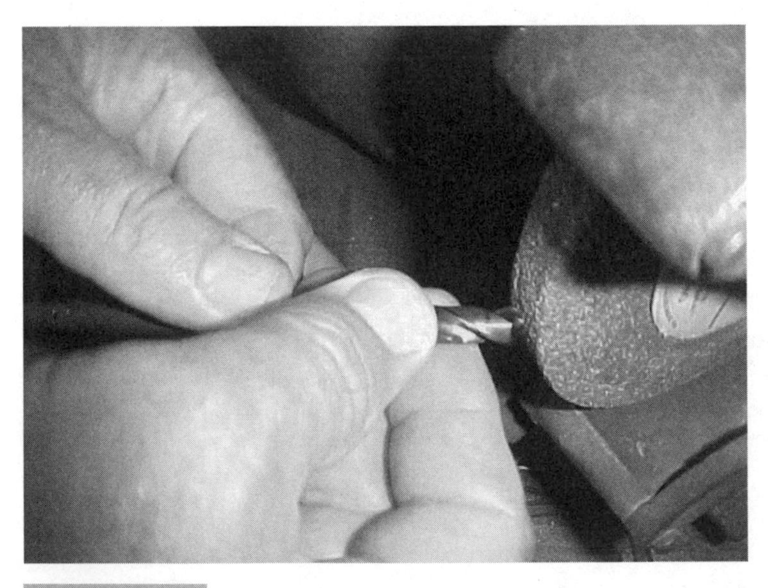

FIGURE 15.6 **Sharpening a drill bit.**

Small Saws

A hack saw is used to cut steel, aluminum, polycarbonate and other stuff. A couple of hack saws are shown in Figure 15.7. A hack saw comes in handy when cutting small parts and material, but make sure you have a couple of new blades handy. When using a hack saw, always apply downward pressure on the away stroke. When pulling the saw back toward you, lighten the pressure. Polycarbonate is really hard to cut with a hack saw because of its small teeth. The material tends to collect in the teeth and cause heating, which makes the material gummy and sticky. Your saw will not like cutting it. So, although you can use a hack saw, I suggest getting either a jig saw with a big toothed blade or a small, upright bandsaw for cutting polycarbonate and other types of plastic.

A jig saw is a small reciprocating saw (shown in Figure 15.8). It's used for cutting steel, aluminum, polycarbonate, and other stuff, depending on the number and size of the teeth on the blade. A jig saw uses a small, vertically aligned blade. Because of this, it's easy to cut circles and arcs. It's also used to rough-cut material to make it easier to handle. Make sure you have the right blade for the material you are cutting; the wrong blade will either break or destroy its teeth. If you want a circle cut on the inside of some material, the easy way is to mark the circle, drill a hole in the material big enough for the jig saw blade to fit inside, and then cut the circle out by hand. If you aren't too good at following the lines, use a file to smooth the edges.

A new breed of rotary cutting tools has been designed to replace the jig saw. Figure 15.9 shows the tool, a circle cutter attachment, and some of the specialized fluted cutting blades. These blades are similar to drill bits, except that the flutes are more elongated. These rotary

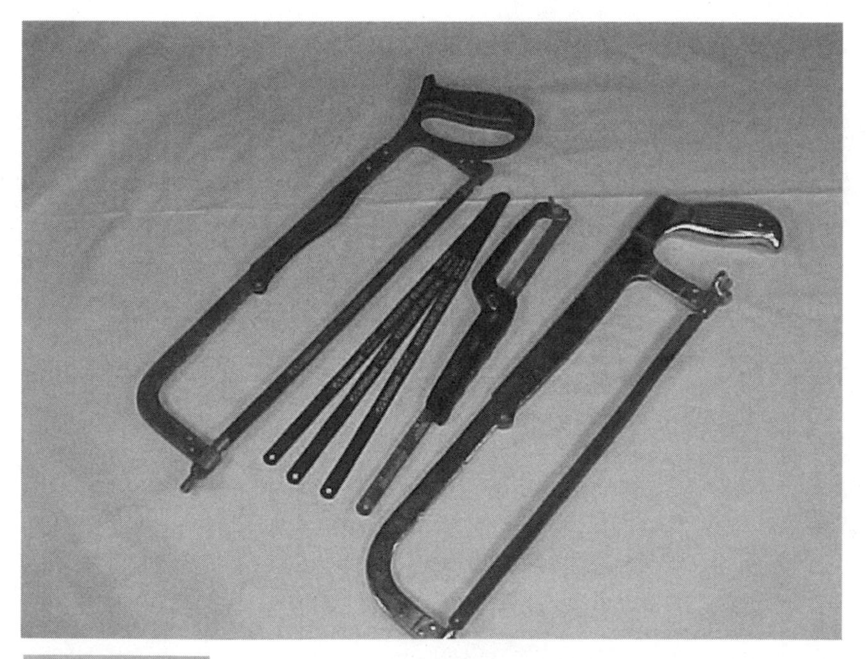

FIGURE 15.7 Assorted hacksaws and blades.

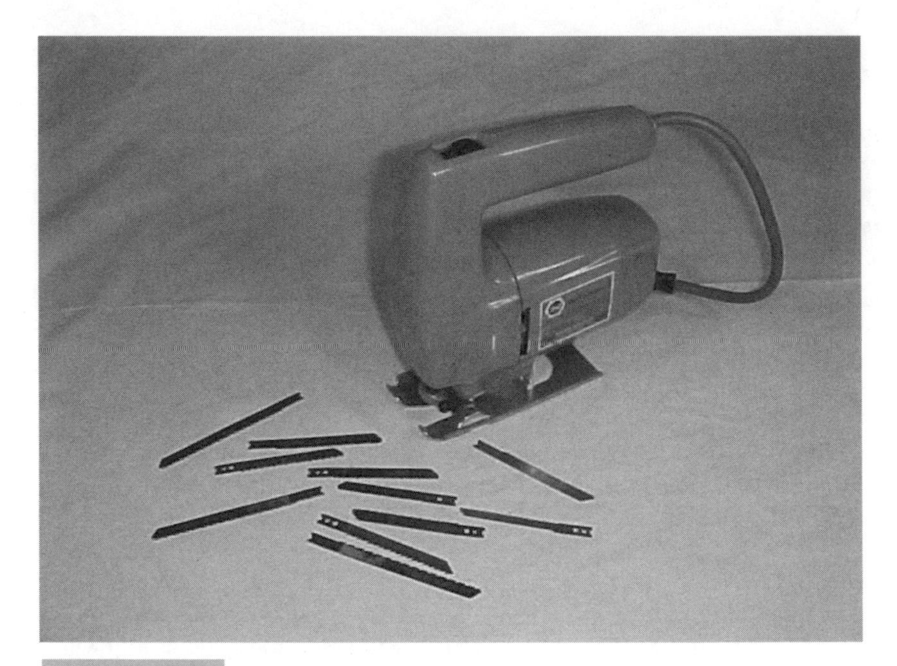

FIGURE 15.8 Jig saw and blades.

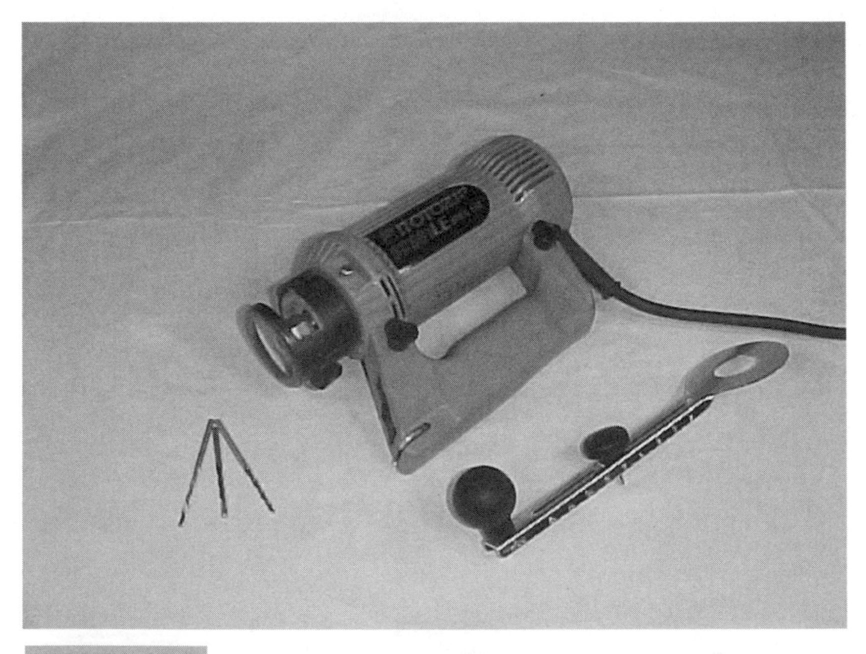

FIGURE 15.9 Roto-Zip, blades, and circle cutter attachment.

cutting tools are pretty nice to have when cutting materials such as wood, polycarbonate, and even thin aluminum. However, they do not like cutting the thick stuff.

Bench Grinder

A bench grinder, shown in Figure 15.10 with a facemask, can be used for an almost unlimited list of tasks. You can sharpen drill bits, form custom parts, clean up cuts or holes, and grind lathe tools. A bench grinder is not only a grinder. You can replace the grinding wheel with cloth or wire wheels for polishing and cleaning parts. Mount the grinder somewhere away from electronic or magnetic equipment, because small metal grindings will be thrown all over the place (unless you have the cash and space to install a vacuum system to pick up these bits on the fly.)

FIGURE 15.10 **Bench grinder and face mask.**

Wire Cutters, Strippers, and Crimpers

Wire *strippers* are really handy, so that you don't have to use a knife while wiring your robot. Some wire strippers have several differently sized, sharp holes along the cutting surface of the tool, and each hole requires a different size of wire. These are designed to wire size standards, so that the wire doesn't get cut when stripping off the insulation. *Crimpers* are used to clamp a terminal on the end of a stripped wire. Crimping tools come in different sizes, so get one that has the crimp surface that you will use most of the time. Most

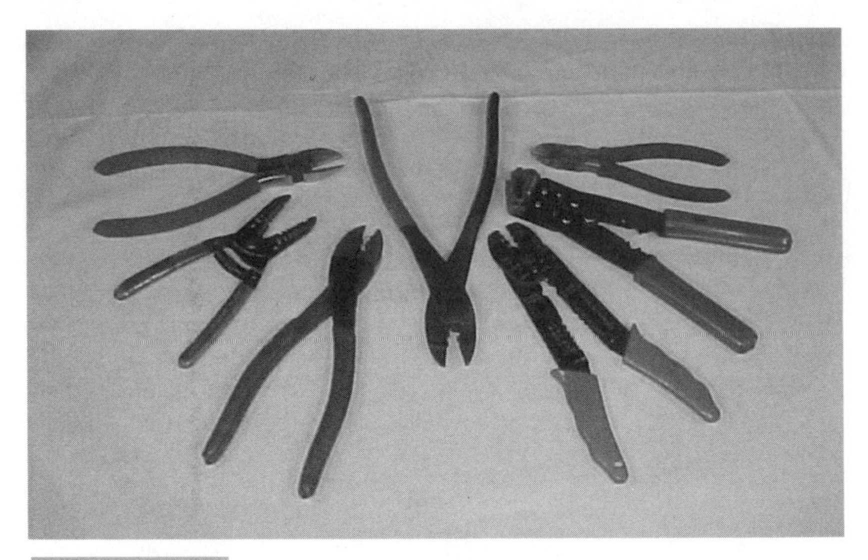

FIGURE 15.11 **Wire cutters, strippers, and crimpers.**

crimpers and strippers come with some part of the tool sharpened for cutting wire. Figure 15.11 shows several different cutter, stripper, and crimper tools. Automatic cutter/strippers cut the wire and then strip the insulation off the end all in one hand motion. I've used the same cutter/stripper set and crimpers for years.

When it comes to crimping a terminal onto the end of a wire, don't try to use a terminal that doesn't fit the wire. Terminals are made for specific ranges of wire sizes and don't work well for others. When closed, some crimpers form a hollowed "U" shape. A terminal has a split down the length of the crimp end. Place the split side into the round bottom of the hollowed "U" shape and squeeze. This ensures that the most metal-to-metal contact is achieved between the wire and the terminal.

Soldering Iron

Unless you are really lucky, you are going to need a soldering iron at some point. Several types are shown in Figure 15.12. Get one that has changeable tips and adjustable heat ranges. There will be times when you need to solder small wires, and you'll need a small tip and low heat. When you have to solder big wires or tabs to NiCd batteries, you'll need a large tip and a lot of heat. Adjustable irons can be expensive, however. The alternative is to get a soldering pencil for the light work and a soldering gun for the heavy stuff. To make a good solder joint, you need the right amount of heat. Cold solder joints look chunky and do not conduct electricity well. Good solder joints look shiny.

One note I should mention is that you should be careful using solder in high-current situations. Parts that are rated for high currents are made with connections that accept terminals and not bare wire and solder. Solder can melt while transferring high currents, causing the wire to let go and break the circuit.

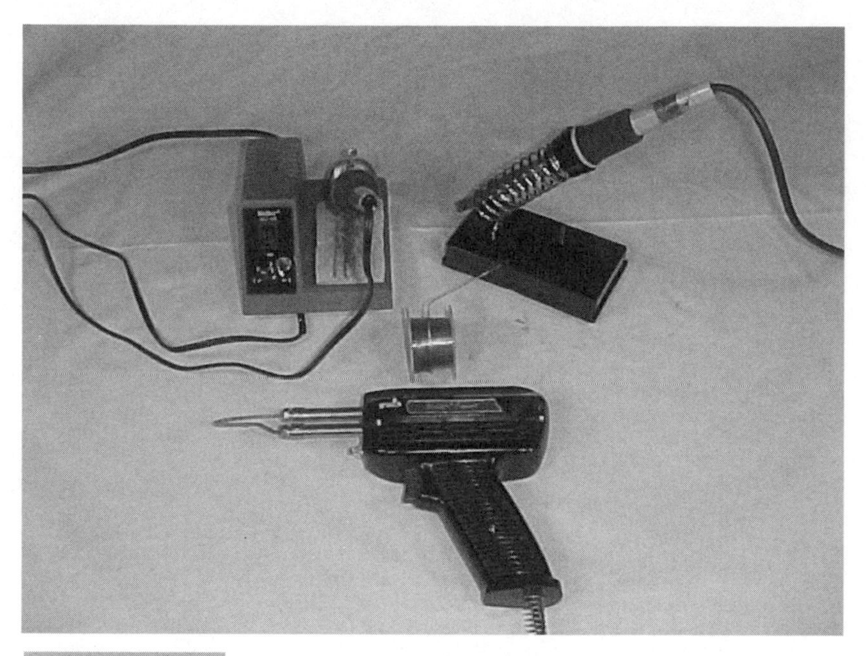

FIGURE 15.12 **Adjustable soldering iron, pencil iron, and soldering gun.**

Wrenches and Socket Sets

You will need a set of wrenches if you ever plan on using bolts and nuts in your robot. Lots of specialty wrenches are available, but I've found that I can't do without a good set of open-end or box-end wrenches with a swiveled box end. Socket sets make it easy to quickly install or remove nuts and bolts. But because I personally like using bolts with cap heads, too, I have a small Allen wrench pack and a set of T-handled wrenches as well. All of these are shown in Figure 15.13.

Micrometer

Micrometers (shown in Figure 15.14 along with a scale) are measuring tools. You can measure the thickness of flat parts, hole sizes, outside diameters, and a number of other lengths to 0.001 accuracy or more, depending on the instrument. Micrometers are a must-have when using a milling machine or a lathe to make parts, but they are nice to have even if you aren't using a mill or lathe. The "mic" I have is the old, dial type. This is fine for me, because I learned how to read one a long time ago. Newer mics have digital readouts. Some mics come as a set along with a 6-inch scale. (*Scale* is a machinist's term for *ruler*.) I use my 6-inch scale much more than any other measuring device.

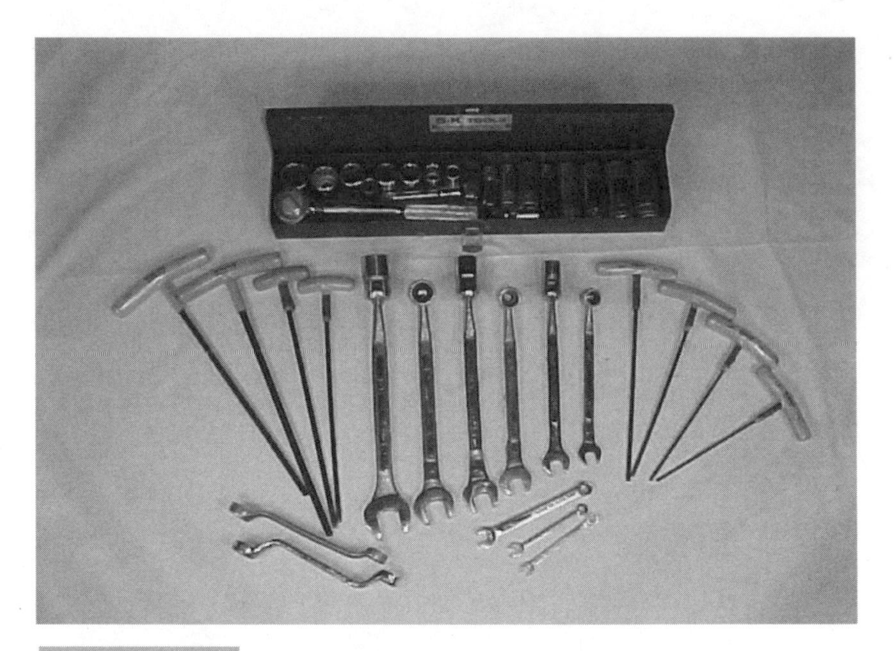

FIGURE 15.13 Socket set, Allen wrenches, and standard wrenches.

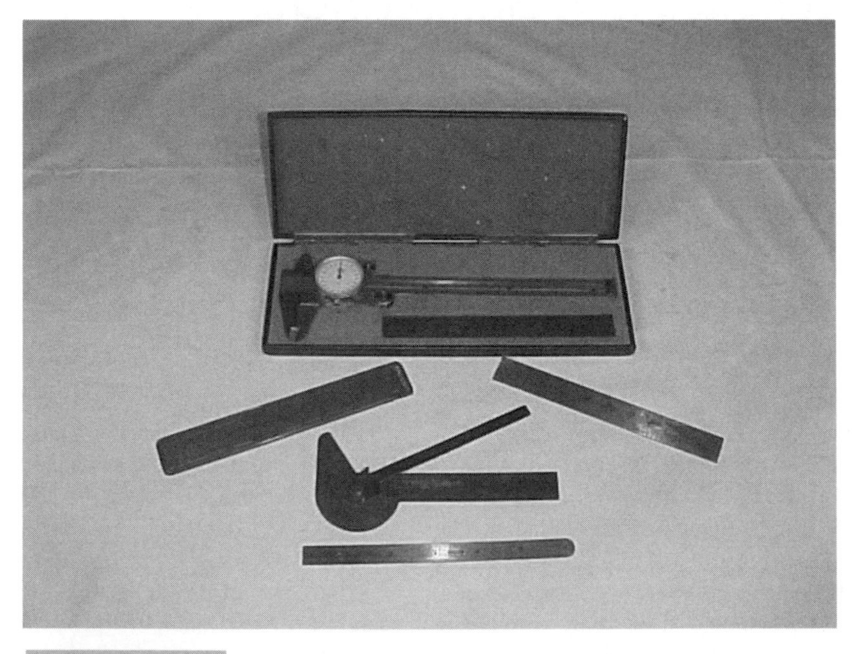

FIGURE 15.14 Micrometer and assorted scales.

Drill Press

A drill press, like that shown in Figure 15.15, makes life easier when drilling holes through round and square tubing. It also makes it easier to drill a hole for a roll pin through the center of shaft material. If you are going to do that, you should pick up a round stock center finder. The *center finder* is simple to use. Put the finder in the drill press and lower the spindle down over the round stock, which is clamped in a vise. Once the marks are lined up, clamp the vise down to the drill press table. Change out the finder with the drill bit and drill your hole. Make sure the drill press table is tightened down or it will throw the hole off center.

FIGURE 15.15 **Drill press.**

Broach Set

A broach set, shown in Figure 15.16, is nice if you plan on using a locking key instead of roll pins. A broach looks like a thin, extremely coarse file with tapered cutting teeth. It comes with bushings that are inserted into the bore of the work piece. The bushing has a slot where the broach is pushed through. Shims are added after a pass, so that the teeth continue to cut. Usually, you use an arbor press or something similar to press the cutter through the work. In the beginning stages of each cut, lift up on the press and let the broach recenter itself to keep the broach going straight. Do this for every couple teeth until you get near the center of the broach. Keep the broach lubricated with lots of cutting fluid.

The other half of this process is cutting a keyway into the shaft to match. This requires either a milling machine or a great deal of patience and a file. I recommend finding someone with a mill or buying some pre-keyed shaft material. Several of the parts catalogs in Appendix E carry pre-keyed shaft material.

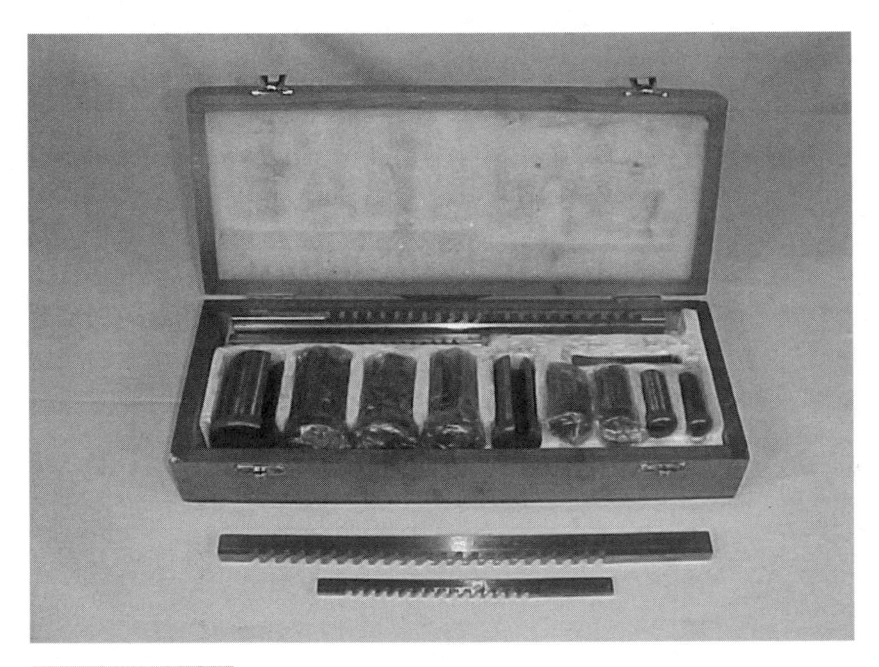

FIGURE 15.16 **Broach set.**

Band Saw

A band saw, like that shown in Figure 15.17, takes the place of a hack saw. It's used for cutting larger materials or for saving your arm from using a hack saw. Band saws have an adjustable vise to handle different sizes of material and to hold it at different angles. Keep the saw blade clear of metal fragments: some blades break or jump off the track at the slightest hint of a fragment. Always support long materials on their opposite ends and in the middle while cutting.

FIGURE 15.17 Band saw.

Angle Grinder

An angle or side grinder, as shown in Figure 15.18, is a handheld power tool used to grind and cut. It's one of those multipurpose devices—many people use it to clean up welds or

FIGURE 15.18 Angle or side grinder.

grind down exposed bolts and I've even used it to quickly open up a robot shell to do repairs when time was at a premium. Once again, you get what you pay for. My machinist friend bought a couple of cheap grinders made in China. Instead of using a ball bearing to support the grinder shaft, the manufacturer replaced it with an aluminum slug of the same size. Just another example of cost cutting interfering with quality.

Welder

A welder is a really handy piece of equipment to have around the shop. My personal welding machine is shown in Figure 15.19. It is small, but it does exactly what I need done. It is not something that you can just pick up and use well without instruction of some kind. A couple of recommended books on welding are listed in the "Other" section of Appendix E. The MIG welder with a CO_2 attachment is probably the easiest to use, although the stick (arc) welder is probably the cheapest to buy. It takes some practice, but using a stick welder produces fine results. In the beginning, it is best if you find or make a friend who has a welder and knows how to use it.

Getting parts welded together in the particular shape you want can be tricky. They key is to set up the materials so that you can move your welding tip in one smooth motion. If necessary, use clamping devises to hold the parts in place. MIG welders usually have an automatic feed on the filler wire. When you pull the trigger, the wire is pushed to the material and the current melts it all together to form the joint. Stick welders are different in that

FIGURE 15.19 Portable MIG welder.

your filler material is used up, and you have to compensate by moving your hand closer to the project. With either type of welder, practice is the key to good-looking, strong welds. Welding causes a great deal of heat, which can deform the steel you are welding in a heartbeat. Never weld long beads without *tacking* the parts together first. Tacking is the act of putting several small welds on the parts to hold them in the correct position.

Punches and Hole Transfers

A punch is a devise used to mark materials with a small dot, dent, or dimple. The dot, dent, or dimple usually marks the spot where you are planning to drill a hole. Besides simply marking the center of a hole, the punch creates an irregularity in the surface so that the drill bit drills where it is supposed to and does not walk. Most punches are made of a solid piece of tool steel. They are meant to be struck lightly with a hammer. Spring-loaded punches can be used without a hammer. You simply place the tip where the dot should be and push down. Once enough pressure is applied, the spring inside the punch releases and causes the mark to be made.

A hole transfer set is another useful tool. Similar to a drill bit index, it has a punchlike instrument for each size of hole. The hole transfer tool is a solid round bar, ground to size, with a sharp point located on the end, which should be directly in the center of the tool. The hole transfer makes it easier to mark the center of an existing hole onto a new piece of material. Figure 15.20 shows some assorted punches and a hole transfer set. In Chapter 21, I use a hole transfer set to mark the holes that exist in the baseplate of the bot onto the side, front, and rear armor.

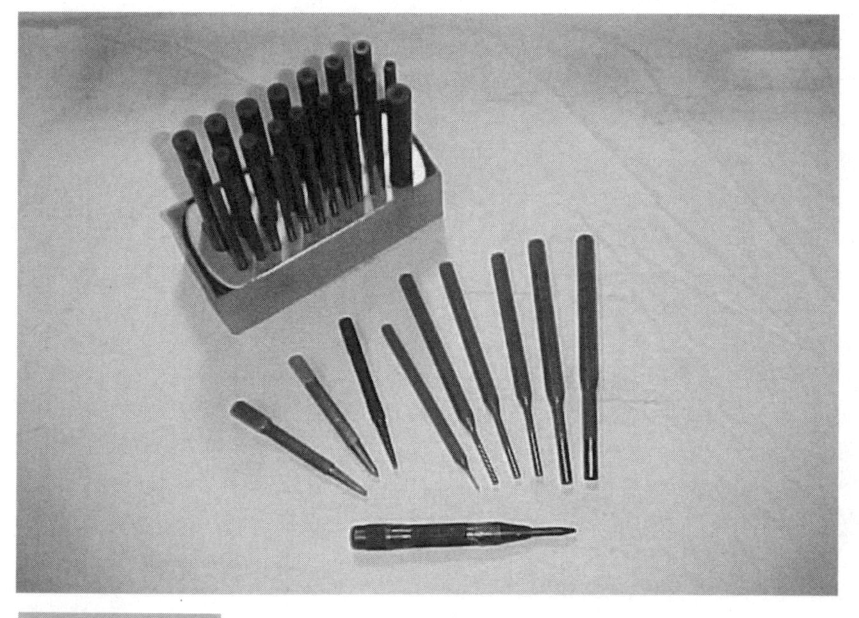

FIGURE 15.20 **Assorted punches and a hole transfer set.**

Mill, Lathe, or Three-Way Machine

A three-way machine combines a drill press, milling machine, and lathe in one. These machines can be pretty handy, but I've heard their quality is a little lower than one would want. Pretty good assembly kits are available, but you supposedly have to modify and improve quite a bit to get it to hold any kind of tolerance. Some people may disagree, so the best thing to do is research and make up your mind for yourself. With the three-way machine, you need a lot of setup time to go from one function to another. If you have the money, you can buy a separate milling machine and lathe, as shown in Figures 15.21 and 15.22. This is the ideal way to go and should be a goal for your future bot building adventures.

Buy a lathe first. It is possible to set up a vise to hold material on a lathe bed and a chuck to hold a cutting tool and perform small amounts of mill work on a lathe. Keep in mind that you may spend as much or more on tooling and accessories as you do for the machine itself. Some of the things you will need or want for a milling machine include a set of collets and end mills, a large vise, an indexer (rotary table for cutting circles),

FIGURE 15.21 Milling machine.

FIGURE 15.22 Lathe.

clamps, and a drill chuck. Some things you will need or want for a lathe include a live center and drill chuck for the tailstock, cutting bits (tools), a few tool holders, and a quick-change tool post. You will need measuring equipment and layout tools if you have either a lathe or a milling machine. This includes calipers, micrometers, dial indicators, scales, and scribes. The more expensive machines have digital readouts that tell you the position of the tool and part. These can make life easier but aren't necessary and do not replace the manual measuring tools. Another luxury for the shade-tree machinist might be a motorized feed. If you own a milling machine without it, I think you will agree.

Summary

This chapter showed the main tools that I use when building a bot. If I had to chose five large tools that a beginner should aspire to buy, I would give you the following ordered list:

1. Drill press and drill index (bits)
2. Bench vise
3. Wrenches
4. Bench grinder
5. Band saw

Tools like the soldering iron, wire cutters, strippers, and crimpers, and micrometers are also necessary from the beginning. A sturdy work bench can be found almost anywhere. I know at least one builder who builds bots in his kitchen. (If you plan to use the kitchen as your workshop, make sure you protect the counter surface.)

Chapter 16 talks about holding things together in a temporary manner. In other words, we start to look at fasteners like nuts and bolts. We talk about the differences between cheap bolts and good bolts, and discuss the proper way to use them so that they have less chance at breaking while in use.

16

FASTENERS

You can join components in three ways: welding, adhesives, and fasteners. Like materials can be welded to form a permanent joint. Unlike materials that need a permanent joint require adhesives. Fasteners hold things together in a temporary state whether they are like or unlike materials. Fasteners give you a way to take pieces apart, whether you plan to do so or not. The main drawback of fasteners is that you must put a hole in the materials you want fastened together.

Bolts and Nuts

The most common type of fastener is the bolt and nut combination. Everyone knows what they are, so I'll skip a description. However, not everyone knows that there are different grades of bolts, some harder and stronger than others. The grade 8 bolt is a favorite among some bot builders; these have a higher tensile strength than standard bolts. Some builders only use bolts approved for aircraft. (Grade 8 is not approved for aircraft use.) Grade 5 bolts have a slightly lower tensile strength and cost a bit less than grade 8. In my opinion, grade 5 bolts are better suited to robot building because of the materials used and the cost of the bolt.

Standard bolts are found at the local hardware or lawn and garden store, which usually carry few, if any, grade 8 or 5 bolts. They are more expensive and there is less of a demand for them. I buy my grade 8 and 5 bolts at a local store that specializes in bolts and other fasteners. (Look in the yellow pages or on the net.) You get bolts cheaper if you buy them by the box, and it's always good to have a healthy supply of bolts in various lengths and thread sizes. Never buy just enough to do what you want: you always need another bolt. Bolt cabinets, sold in various stores, have 1,100 or more bolts, nuts, and washers. These can be a good starting point but they do not usually have the graded bolts and always contain a lot of things you probably won't use. I suggest buying an empty cabinet and stocking it with several of the common sized, graded bolts, nuts, and washers. My collection consists mainly of quarter-twenty threaded, socket head bolts in varying lengths from 3/4 inch up to 2 1/2 inches. I also have some 5/16 threaded socket bolts, quarter-twenty-eight threaded, and some 3/8 eighteen bolts. I also have boxes of nuts, flat washers, and split washers for each size. It is very nice to be able to grab the exact bolt I need instead of scrounging around for something that might work. Figure 16.1 shows a few of the bolts I normally stock.

BOLT MARKINGS

Some bolts have markings on their heads to let you know what type they are. Figure 16.2 shows some grade 8 hex-head bolts. The number of lines on the head of a bolt determine the grade. Refer to the table of "Bolt Grade, Tensile Strength, Torque Specification and Shear Strength" in Appendix C. Grade 2 bolts, the kind you find at local hardware or lawn and garden shops, have no lines. Grade 5 bolts have three lines. Grade 8 bolts have six lines. I like to use black, socket-head bolts, somewhere between grades 5 and 8. Other builders like stainless steel bolts. The only real advantage of stainless bolts is their resistance to rusting. They are primarily made for the food service and marine industries; they cost more and are not particularly stronger than the grade 5, grade 8, or black socket-head bolts.

FIGURE 16.1 Socket head bolts.

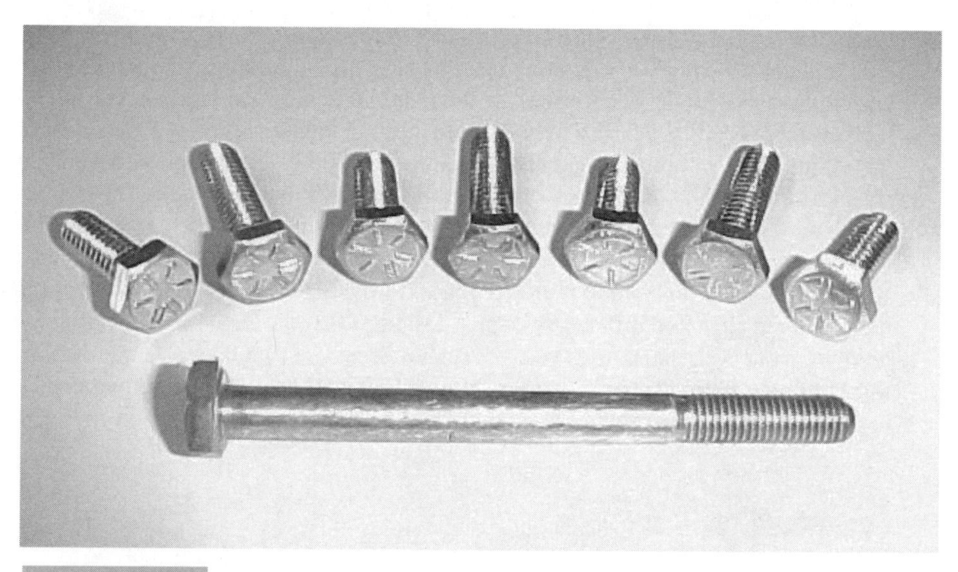

FIGURE 16.2 Grade 8 hex-head bolts.

ORDERING BOLTS

If you use them enough, you eventually learn to recognize which bolt is which with regard to threads and sizes. When ordering or asking for bolts, you must know how to ask for them. If you are bolting something together that has a 1/4-inch diameter bolt hole, you need a bolt with a 1/4-inch diameter. You also need to know which thread count you want. The common 1/4-inch diameter bolt thread count is 20 or 28. I mostly use 20—that's 20 threads per inch. You need to know how long a bolt you need. The length of a bolt is usually measured from the bottom side of the head to the tip of the threaded end and expressed in inches. If you need a bolt that is 1-inch long, you ask for a "quarter-twenty, inch-long bolt." If you want a fine thread, ask for a quarter-twenty-eight, inch-long bolt.

Many bolts longer than about an inch are not threaded all the way to the head. The non-threaded part is called the *shoulder*. You can get bolts that are threaded all the way up; these are called *tap-through* or *fully threaded* bolts. The use of these depends on the application, but the idea is that you don't have the material that is being fastened together in contact with the threads of the bolt, which usually makes for a stronger joint. I don't stock the fully threaded bolts in large numbers.

PROPER BOLT USAGE

Unless you mount and tighten it correctly, the grade of the bolt does not make any difference to whether it breaks while in use. Every bolt grade and size has a torque specification (consult the table of "Bolt Grade, Tensile Strength, Torque Specification, and Shear Strength" in Appendix C). This specification can change depending on what the bolt is holding together or tightening against. For instance, if you are bolting two pieces of polycarbonate together, the material deforms and possibly breaks before you reach the ideal torque for the bolt. In cases like this, it is advisable to use some type of thread locking solution or a self-locking nut so that you can be sure the bolt does not vibrate loose. Washers help to spread the load across more of the material being held together.

Bolts are meant to be used to hold materials together. This concept is sort of obvious if you think about it, but many people do not realize that it is the friction between the materials created by the bolt pressing on them that makes the connection. The bolt should not be depended on to bear any twisting load on the joint. Figure 16.3 shows bolts being used properly and bolts that have been put in *shear* or in a position that can cause them to break.

The illustrated bolts are in the same places in the materials, but the forces applied to the materials are absorbed differently. Better design of the joint may or may not help the bolts hold the materials. Sometimes you just cannot design so that there are no bolts in shear. In this case, you should at least use more bolts. The best thing to do is use hardened steel pins to take the side forces and bolts to hold the materials together, as shown in Figure 16.4.

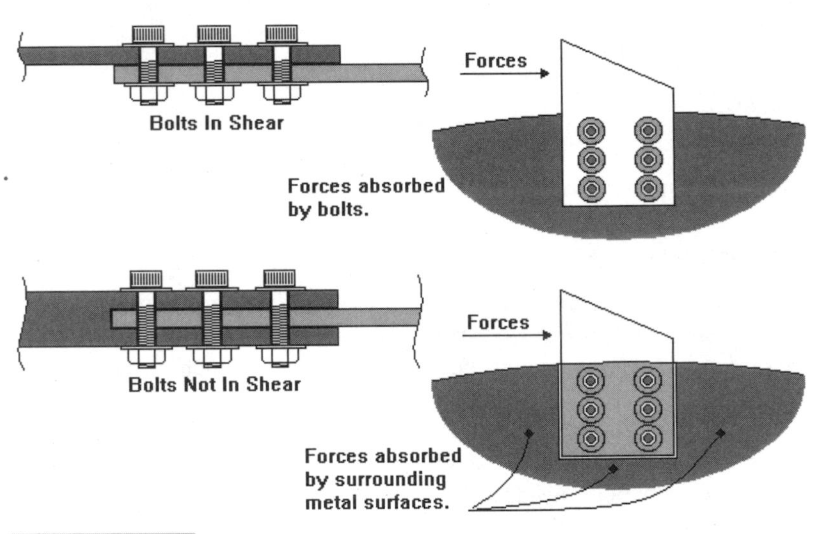

FIGURE 16.3 Bolts in shear and in proper use.

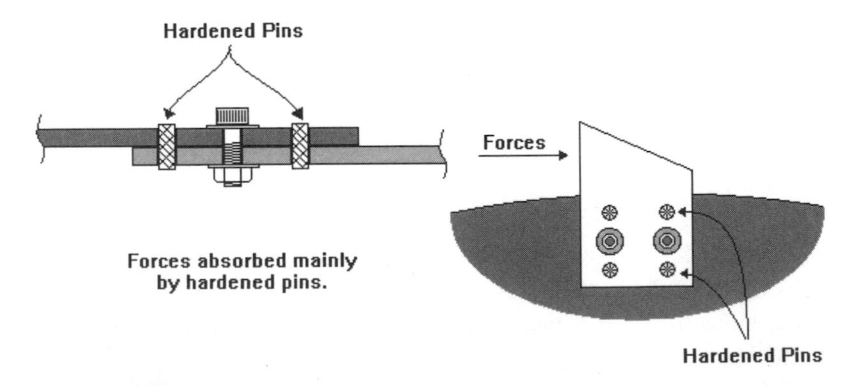

FIGURE 16.4 Using bolts and pins together.

SMALLER BOLTS

Machine screws are simply small bolts with a screwdriver-type head, as shown in Figure 16.5. They, too, come in different sizes and threads. I keep a small number of these around, because they are useful when it comes to mounting electronic boards and such.

Sheet metal screws have large heads and a sharp spiral thread. The thread is much larger than that of a machine screw or a bolt. These screws are good for mounting light armor or fastening things to light armor or other thin materials. Three different types of sheet metal screws that come in handy are shown in Figure 16.6. The *self-drilling screws* are on the left. Instead of a point with threads on it, the self-drilling screws have a tip similar to a drill point. A *self-drilling, self-tapping screw* drills its own hole and makes its own threads. The *self-tapping screw* makes its own threads, but you must drill the hole. Each of these can be removed and reused on other materials unless significant damage has occurred.

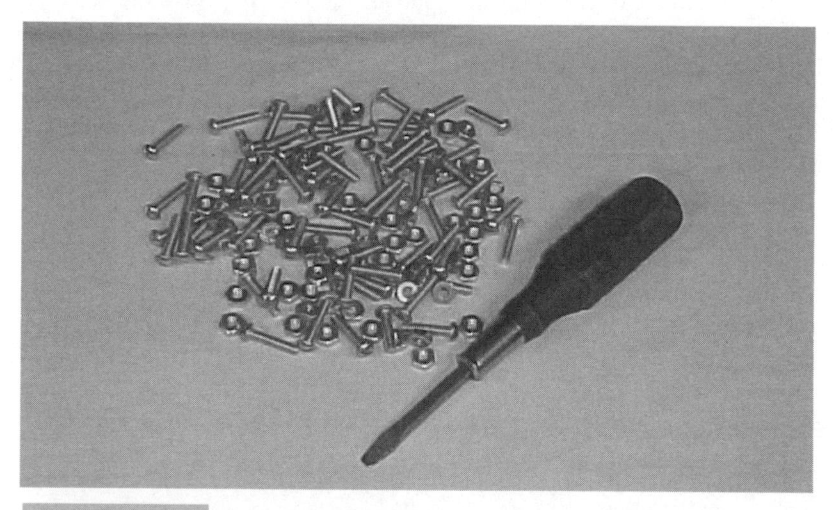

FIGURE 16.5 Machine screws.

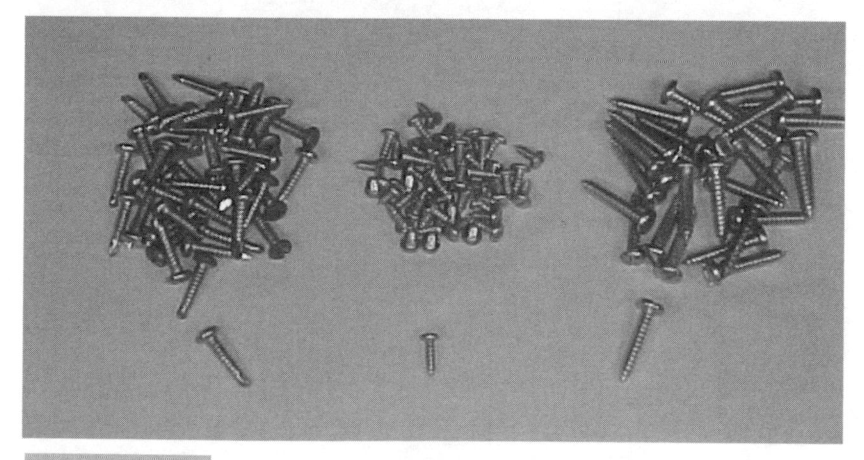

FIGURE 16.6 Sheet metal and self-tapping screws.

Rivets

Pop rivets are really useful when you can't reach the other side of what you are fastening together. They have a big drawback though: to get back into something you have riveted shut, you have to drill the rivets out, which can leave small metal bits inside your robot. Avoid rivets if you can but just in case you can't… Rivets work by expanding inside the hole you provide and holding the two pieces of material together. They are not made to withstand large shearing forces, so do not depend on them to do so.

Rivets, which look like a nail with an upside down head, need a special rivet tool to set them (see Figure 16.7). The rivet tool grips the nail part or *mandrel*. The upside down head part or *body*, goes in the hole in the material. You then pull the tool's trigger or lever arm to squeeze the rivet head down. This takes a few pulls. Once the body has compressed and expanded out toward the edges of the hole, the mandrel starts to take the strain of the tool. The mandrel is made to break at a specific amount of pressure. When that happens, you hear a loud pop and the body of the rivet is left holding the material together while the rivet tool still holds the mandrel. Figure 16.8 shows two pieces of metal being held together by a rivet. It also shows the broken mandrel, still inside the rivet tool. Be sure the tool has the right size nose to hold the mandrel. Also be sure the hole in the material is the right size for the body of the rivet. Those two specifications are usually given by the manufacturer of the tool or the rivet.

Rivet nuts are a bit different from the standard rivet but work pretty much the same way. Rivet nuts can make threads in material that is too thin to be threaded with a tap, because they have an expanding body that tightens on the existing hole. Once the rivet nut is

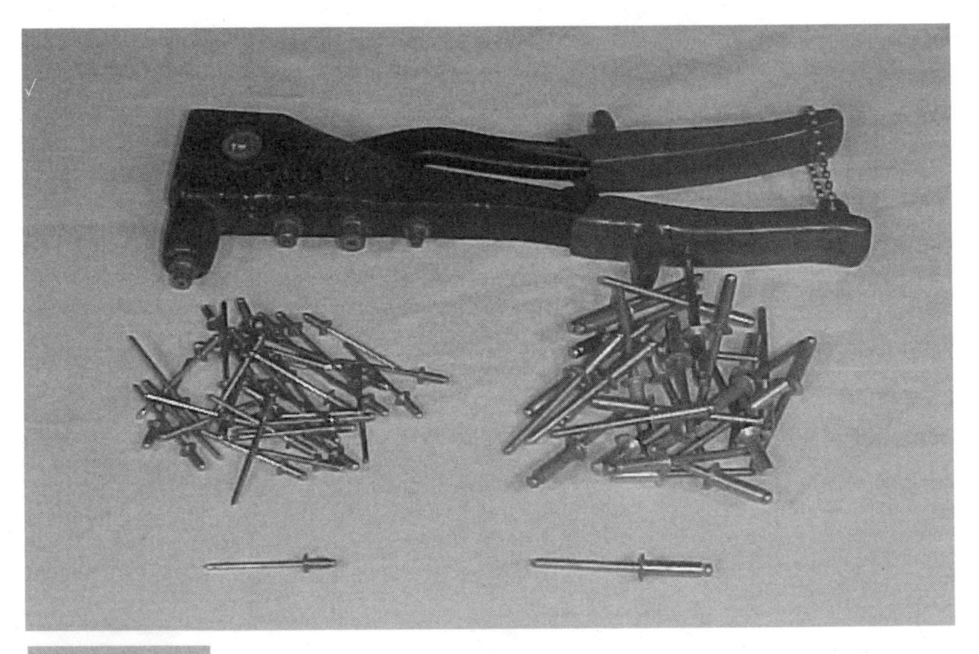

FIGURE 16.7 Rivets and a rivet installation tool.

FIGURE 16.8 Installing a rivet.

installed, you can screw a bolt into it to secure what you want. To install a rivet nut, drill the appropriate size hole in the material. You may have to tap the rivet nut into position with a hammer. If so, put a bolt inside the nut so that the threads are not bent in the process. Once the rivet nut is in the hole, put a nut on the exposed end of the bolt. Tighten the nut and bolt until the rivet nut body is compressed and holding against the material. Take the bolt out and you are ready to set the next one.

Weld Nuts

I've welded regular nuts to a piece of angle iron so that I could install and back out a bolt without holding the nut. This is a pain to do because I usually get weld splatter on the threads of the nut or heat it up too much and deform the threads. A weld nut solves this problem. They come in many different shapes, but the one I use is flared out on one end (see Figure 16.9). To install them, you drill the hole big enough to accommodate the body of the weld nut. Then insert the nut into the hole and weld the flared end to the material. These work wonders when you need to seal off the last piece of armor and it is held on by bolts. Imagine bolting the sides of a six-sided box together: you can get to all the nuts and bolts until you get to the last piece, so it better have weld nuts in it.

FIGURE 16.9 Weld nuts.

Summary

Fasteners are a very large subject, so large that we can't do it justice in this book. What we've covered here are the very basics: simplified bolt markings, how to order bolts, and how not to break them by using them incorrectly. Rivets can be your friend if you have some small pieces to hold together, but they should not be trusted with large loads. Weld nuts just make life easier.

In Chapter 17, we talk a bit about where you can find robot parts. The Internet is a great "buyer beware" parts store. Junkyards can contain a gold mine of parts and materials. With respect to robot parts, the old saying "One man's trash is another man's treasure" can hold a lot of truth. We also talk about how to get free stuff and even money from sponsors!

17

GETTING PARTS AND SPONSORS

I must get at least one email a week asking where to get parts for a bot. If all the veteran builders are getting this same amount of email, there are a lot of people who need to read this section. Robot parts are literally everywhere. Some people build robots exclusively from junkyard parts. Others make or have made, custom parts. Everyone finds parts in different places.

The Internet

The easiest place for me to find parts is on the Internet. Hundreds of online vendors sell metals, electronics, wheels, saw blades, and anything else you can think of. In fact, most builders I know buy all their speed controllers on the Internet. Lots of parts can be found at online auctions. I bought my first remote control unit in at online auction. Just remember, *buyer beware*.

Builders are also getting into the parts business. Some sell parts off robots they do not plan on using any more, others create partnerships with known vendors to try and corner the bot builder niche market. Some design and build their own line of custom parts. None of these sources is more than a click away in a search engine. The phrase "killer robots" brings up at least three sites to search (try the name of your favorite robot TV show).

Surplus

Across the country, thousands of junkyards, surplus houses, and flea markets are gold mines of robot parts. Stop everywhere you see scrap metal piled up on the fence. See if you can look around and possibly buy something. Many of these establishments even have Web sites with limited catalogs of merchandise. And some of those will even keep an eye out for a certain part you might need. Check the "Parts Catalogs" section in Appendix E. Call the company or visit its web site and ask for a catalog. You can sometimes get old, but useful, catalogs from local industrial parts distributors.

Sponsorships

Every bot builder I know would love to be paid to build his robots. This is not the case for most of us; however, there is a possibility of getting a sponsor to donate parts, services, or even money to your cause. But do not expect a company to just hand over $40,000 so that you can build the bot and compete. They need something in return for the money. That something usually involves getting the name of the company on television, but it also includes making them part of the team. Sponsors do not usually get to control what the bot looks like or how it works, but they do want to feel involved. That can mean you will do a couple live demonstrations, which are always fun.

The most common, and easily obtainable, form of sponsorship is the discounted or free part or service. Companies can usually deal with one of their guys cutting up a piece of steel, turning an axle on the lathe, or giving you an expensive part at their cost. All you need to do is work for it. You should not just walk into the store or shop and start asking for freebies. First, make sure it is a store or shop that has a few things you need. Talk to the person selling the part or service and let them know what you are doing. Show them drawings, pictures, or even the unfinished bot. Showing them a working bot is even better. Ask their opinion on how to do whatever it is you need done. Lots of people have opinions on how they would do things. Not all opinions are created equal, but listen to all of them, smile and nod. Tell them you will put it on the list of things to discuss with the rest of the team. Overall, make friends with them. Go back to the store even if you need nothing else from them. One of these days you will. If you kept them up to date with your project, they will remember you next time and might help you out a little more.

All the above works well for local shops and stores, but a lot of vendors on the Internet would like to help out budding teams. More of these vendors want to help out an experienced team simply because there is a better chance that they will have a working robot to get the company a little more exposure.

Having a Web site helps you establish a presence in the world. Do not just grab any free Web hosting service and put up a couple of flashy team graphics with a 3D CAD drawing of a bot. Your Web site should have clear pictures of the actual robot and any competitions that you have attended. The more content, the better—after all, on what information will the potential sponsor base their judgment of you and your team if you are not local to them? Contacting nonlocal sponsors is a hit-and-miss proposition. You must get their attention with the first page, whether its a Web site or a brochure. If they are the slightest bit interested, they will read the rest of what you have sent. If not, they will toss it into file 13.

Once you have someone interested, you need to clearly outline exactly what is expected of them and what they can expect of you. Write out a contract, if necessary, but at the very least you should discuss all the possibilities with them. Make sure you feel you are getting a fair deal. To do that, you must do your research. Find out the potential benefits of being a sponsor and do your best to make them pay off for the company. Make sure the company knows the limits of what you can offer. Negotiate with them. They are interested because this is a fun sport that a lot of people are getting into. Keep it fun for them by keeping them involved.

Summary

Finding standard parts should be one of the easier tasks in robot building. Remember to search the Internet, but do not believe exuberant claims. Make smart purchases from reputable sellers. Check out junkyards and surplus outlets. I have found many motors and other parts just by stopping in for a 30-minute walk-around. Last, give your sponsors the feeling that they are involved in your team. Keep them informed and ask their advice.

Chapter 18 is the last "theory" chapter before actually building a fighting robot. In it, we talk about going to competitions. You need to get a feel for it before you jump right in, and you also need to do some preparation so that you can fully enjoy the competition.

18

COMPETING

Next to building your robot, going to competitions has to be the best pastime ever. As of right now, only two BattleBots™ competitions are held each year, and a few TV shows film their own competitions as well. Several private competitions are held, too. Probably the most notable is the BotBash™ (www.botbash.com) in Phoenix, Arizona and the North Carolina Robot StreetFight™ (www.ncrsf.com) that I organize and produce. More competitions pop up all the time. Eventually, with growth of the sport, there may be regional events. Small weekly events may even be a reality a little further down the road. When competing concentrate on three issues: the budget requirements, tools, and life in the pits. If you do all these right, you will have a most enjoyable experience.

Budget Requirements

You have already spent a great deal of money, either yours or a sponsor's, by building the metal eating monster sitting in your garage. Now you get to spend even more money on airfare, hotels, rental cars, shipping, food, entertainment, emergency parts, and souvenirs. When you first planned the building of your bot, you should have included money for all these things within the budget. You may be able to do without entertainment other than the competition, souvenirs, and even food for a week but you have to be there and your bot will need repairs. Hotel costs vary depending on the location of the event and the time of year. In the past, veteran builders were given free hotel rooms for the duration of certain events. I do not think this practice will continue in the future, due to the number of new entrants at each competition. However, the event promoter may be able to get a sponsor to pay for it. Those few events that are "invitation only" usually pay for the hotel, transportation, shipping, sometimes food, and sometimes they pay an appearance fee. The smaller events like BotBash and NC Robot StreetFight offer an opportunity to travel shorter distances, stay for shorter amounts of time, and spend less money. However, the robot action is every bit as exciting and destructive as what you see at the big show.

It's not rare to get to an event and remember something you need but did not bring. The least of these is a camera and film; the worst would probably be a speed controller, motor, battery, or RC unit. (Good luck finding one of those at short notice but do ask around the pits.) Either way, you will need some spending cash included in the budget. The best souvenirs I can think of are T-shirts from other teams or the event itself. Most teams that have T-shirts sell them at reasonable prices. Some may be willing to trade for a cool team T-shirt of your own. If you have not done it yet, you might want to look into getting shirts, hats, and/or stickers made up. They get your team name and sponsors some recognition and may get you a free shirt as well.

Tools in the Pit Area

There are a few things that you should do when carrying and using a load of tools in a competition halfway across the country. For starters, mark your tools with paint or colored tape to help you identify your tools if they get mixed in with others. Someone always needs to borrow some tool or other, and so far,\everyone has been really good about letting anyone borrow tools. Some even go so far as to say "take anything you need, even if I'm not here, as long as you return it." The key statement is the "return it" part. Return any tools you borrow in as good or better shape than when you borrowed them. Make the lender happy he said yes when you asked. Once done with a tool, return it immediately. Not only will you forget to take it back an hour later, the lender might need it now. Above all, get to know the people beside you. You will make friends, and they will watch out for your stuff while you are gone. You'll do the same for them.

Life in the Pits

While in the pits, keep all electric tools unplugged when not in use and keep all sharp things, on the bot and in the pit area, covered. Electric tools can go haywire if something is set down on them or they are kicked around in a pit area full of people rushing to get their bot, repaired for the next match. Keep your safety and the safety of people around you in mind. Pit areas are typically very crowded as it is. Try to keep your area as clean and clear as possible. The bottom line is *safety*.

Summary

I am sure you are ready to jump in and build a bot, and in Chapters 19, 20, and 21 I show you how to do just that. I decided to include three projects. Chapter 19, Project One, RAID, relies on your creativity and scrounging abilities. It will get you started really quickly with an ant weight (1 pound) robot that you can build from things around the house and parts that come with a new radio control setup. Chapter 20, Project Two, Six Million Dollar Mouse, outlines the build process I took when I constructed a heavyweight (220 pounds) robot to compete in the first televised version of Robot Wars Extreme Warriors in June of 2001. It is not meant to show you how to build that particular robot, but it shows what building a big robot entails. Chapter 21, Project Three, Dagoth, is named for one of the gods in the *Conan* (the Barbarian) novels. Dagoth is a light (30 pounds) robot that can be built for between $1,000 and $1,500, depending on your shopping skills. We will construct Dagoth step by step.

After Chapter 21, we'll talk about practice and weight control. I have included several appendices that I think will be useful. In addition to information tables, I offer you a list of basic bot building Do's and Don'ts and a lot of conversion factors. There is also a huge FAQ section containing nearly 60 of the most frequently asked questions of first-time bot builders on the BattleBots™ builders forum on DelphiForums. Don't forget the useful Web sites list and the long list of further reading materials. Now let's get started building robots!

19

PROJECT ONE: RAID ($100)

RAID, named for the bug and ant pesticide and designed after one of my favorite big robots, Chris Hirsch's 911 Super Heavy (www.911superheavy.com), is a 1-pound robot. The main ingredient in this small but fun robot is the remote control setup that you will need for any robot. You can buy a cheap remote setup for around $100, but you are better off spending about $300 on a nicer setup so that you can use it on other robots. I constructed RAID from junk I had lying around the house, so it will probably be difficult for you to copy exactly what I built. I like that idea though. The purpose of the book is to gain the knowledge needed to build these robots and then put that knowledge to use along with your own creativity, not do everything I do. Have fun with the project and make it your own.

Modifying Servos

Servos, shown in Figure 19.1, usually come with most remote control setups. If the remote system you buy does not include servos, you can pick up a couple of cheap ones at your local hobby shop. Servos are small gearmotors that accept commands directly from the remote control receiver. They are primarily used in model airplanes and cars.

Servos typically have three color-coded wires. The colors are usually red, black, and white. Sometimes, depending on the brand of remote they are meant to match up with, other colors are used, but there are always only three wires. A "Chart of Servo Lead Colors

FIGURE 19.1 New servos.

by Manufacturer" is included in Appendix C. For this robot, my servos have red, brown, and orange leads. Red is the positive voltage supply line. Brown is the ground line. Orange is the signal line. Check the chart and your remote control owner's manual to make sure you are connecting the servos to the receiver correctly.

Servos are unlike normal gearmotors in that they do not rotate for a full 360 degrees and they have the remote control interface built in. They only rotate about two-thirds, or less, of a full revolution, because there are mechanical stops in the housing of the servo and a stub on the output gear. When the stub meets a stop, the servo stops turning. The reason for this is that the electronics inside the servo include a potentiometer, or variable resistor, to detect the angle of rotation of the output shaft. The potentiometer cannot be rotated 360 degrees without being destroyed. You have to make some small modifications before you can use servos as drive motors. To open a servo, remove the four long screws holding it together. If there is a disk or arm (control horn) attached to the servo drive shaft, remove that first. Next, take a look at Figure 19.2. See the large gear on the left with a stub that is colored black? I colored it black so it would be more visible in the picture. Use a sharp knife or Dremel tool to fully remove the stub on your gear. Do not damage the teeth of the gear.

Once you lift the gear with the plastic stub you notice a metal post under it. Figure 19.3 shows a brass ring with a plastic clip on top. This arrangement differs among manufactures, so your servos may not appear the same way, although they should be similar. In the center of the brass ring is the metal post, which is the shaft of the potentiometer for the

FIGURE 19.2 Servo gears with stub marked.

FIGURE 19.3 Brass ring with plastic clip.

servo electronics. For this particular brand of servo, the only step left is to remove that small plastic clip from the brass ring.

You must make sure the potentiometer is centered. Turn it to the right and to the left, to its physical limits, to figure out where the center is. In other servos, pull the circuit board from the housing and push the potentiometer out as well. Small plastic tabs may hold it in. Some servos even have screws under the large gear on the right, which hold the motor in the housing. If your servo has these screws, they must be removed to take the servo apart completely. Figure 19.4 shows what a standard servo electronics board looks like.

Many people remove the potentiometer completely and replace it with two standard 2.2k resistors, available from Radio Shack. I'm usually in a bigger hurry than that and just cut the tab off the potentiometer. If you do this, you must remember to center the servo before gluing it. To zero the servo, plug it into your receiver while it is apart. Hook up the battery and turn on the transmitter. Center the potentiometer so that there is no motor movement in the servo. Glue the shaft into that position after cutting it.

Figure 19.5 shows how to replace the potentiometer with the two standard resistors if you would like to do the job that way.

FIGURE 19.4 Servo electronics board.

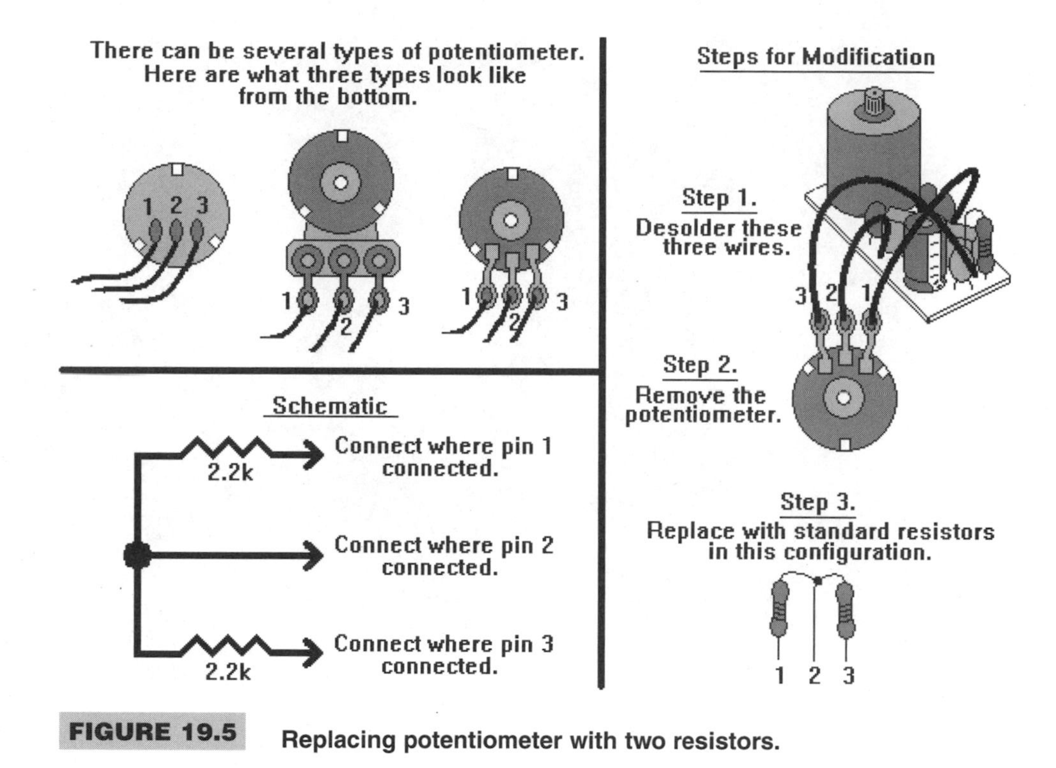

FIGURE 19.5 Replacing potentiometer with two resistors.

Putting Parts Together

After you reassemble the now modified servos, you need to attach the wheels. I used the wheels (shown in Figure 19.6) from an old remote-controlled car toy I had lying around. Since the drive shafts of the servos are plastic with a threaded hole in the center, I was able to use the same small bolts to mount the wheels to the servos. I also put a drop of superglue on the joint to ensure that the wheels didn't spin independent of the drive shaft if the bolt vibrated loose. You can find suitable wheels at your local or Internet hobby shop. Some wheels cannot be mounted directly to the output shaft of the servo; rather, you must use one of the control horns that came with the servo. It will slip onto the output shaft and be held on by a small screw. The wheel should be screwed onto the control horn. Make sure you center the wheel. If you don't, the robot will wobble when it rolls.

Next, I glued the two servos together back to back. This forms the whole of the drivetrain. Then I glued both servos to 1-inch wide strip of aluminum, as shown in Figure 19.7. You may want to use a small clamp to hold everything in place while the glue dries.

The strip of aluminum was just about a foot long. After the glue holding the servos had dried, I put a bend in the strip. This formed the rear support of the robot and gave me a place to mount a weapon motor (see Figure 19.8). The weapon motor came from the same remote-controlled car toy and had a thin aluminum bracket. I used this bracket to mount the motor on the strip. I used superglue and a bit of solder to hold the bracket to the strip. You must have a really hot soldering iron to get the solder to stick, because the aluminum

FIGURE 19.6 Modified servos with wheels attached.

FIGURE 19.7 Servos mounted on the base.

FIGURE 19.8 **Spinner motor mounted.**

carries the heat away really quickly. An alternative is to drill a couple of holes and use rivets. In retrospect, the rivets are probably the best way to do this.

Next, we must mount the receiver and battery pack shown in Figure 19.9. I placed the battery pack inside the bend of the aluminum strip. This was as close to being "on top of the wheels for more traction" as I could get it, because the spinner motor stood in the way of getting it any closer. The battery pack is the same one that came with the receiver. If you get serious about building ant weight robots, do some research into smaller, lighter batteries and better drive motors.

The receiver and transmitter setup that I used are really old. I picked them up from eBay a while back for testing purposes. On this robot, the weight limit was creeping up on me pretty quickly. I decided to save an ounce or so by taking the shell off the receiver. This isn't recommended, especially in the configuration I've given, because it is very vulnerable to side attacks now. I placed the receiver directly on top of the battery pack, as shown in Figure 19.10, and secured it there with double-sided tape. If you do this, make sure none of the receiver's circuit boards are touching any metal. The antenna is wrapped up and taped to the underside of the top of the aluminum strip. Normally, you would not do this but you can get away with it on ant robots because you are standing only a few feet away when operating them.

The two drive servos are plugged into channel 1 and channel 3 of the receiver. This gives two-stick control, as I outlined in Figure 2.2 in Chapter 2. The spinner motor is turned on and off directly with a Team Delta R/C Solid State D-Switch controlled by the receiver's channel 5. Channel 5 is usually the landing gear switch and is usually a simple on–off mechanism. I used a 20-volt, 9-amp version. Measure the stall current of the

FIGURE 19.9 Batteries and receiver.

FIGURE 19.10 Batteries and receiver mounted.

motor(s) you will use and make sure the D-Switch can handle it. (For these ant robots, the motors usually aren't big enough to require 9 amps.) Figure 19.11 shows the complete electrical diagram of RAID. It is very simple to connect: I turn the power on and off simply by disconnecting the battery from the receiver. No switch necessary, but some competition rules may disagree. If you plan on competing with your ant robot, you should check the rules of the competition to make sure that is legal. The last thing I did was add the spinner blade, as shown in Figure 19.12. Be careful when doing this. You must secure the blade properly—if you do not, it will come off and hit someone when you power the spinner motor. I roughed up the end of the motor shaft, drilled a hole in the blade, and then put the blade on the shaft and soldered both sides.

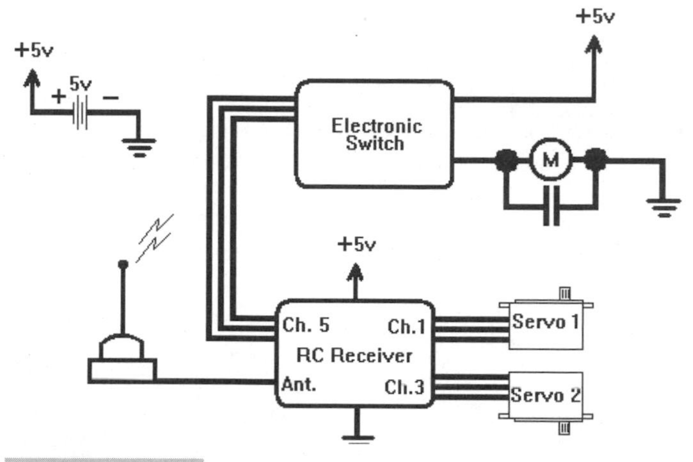

FIGURE 19.11 RAID's electronic system.

FIGURE 19.12 TD switch mounted along with spinner blade.

One last thing I'll mention is that sometimes you need to reverse the way the servo spins when you move the transmitter stick a certain way. This is apparent when you push the stick forward and the servo spins in reverse (relative to the robot). Every radio system I have seen has channel reversing capability. Many times, it's a matter of changing the position of a dip switch on the transmitter, but sometimes you have to program the transmitter. Either way, check your radio manual if you find you need to reverse the spin direction. If your radio does not have this capability, you are in for a little work. You will need to remove the cover of the servo again. This time you will unsolder the motor from the circuit board. Turn the motor around and solder it in the opposite way. This should only be necessary on one servo, if at all.

Cost Rundown

That is all there is to RAID. You may want to build something lighter initially so that you can add some armor later. A quick rundown of what I spent on this robot follows:

Kyosho 5 Channel PCM Radio from Ebay	$65.00 (with servos and batteries)
Team Delta RCE200(C)	$24.50
Old Remote Controlled Toy from yard sale	$ 5.00
Superglue	$ 2.00
Aluminum strips	$ 0.00 (junkbox goodies)
Total	**$96.50**

Add $1 if you replace the potentiometer with resistors.

Add $185 to replace the Kyosho with a new, 6-channel Futaba 6XAPS PCM radio with servos. (Total $281.50.)

Summary

Building RAID should be the easiest project in the book, because the whole robot relies on your creativity and scrounging skills. The electronics of the bot are not complicated. Modifying a servo is easy but it takes patience. Depending on things you have lying around the house, you can get away with an extremely small budget and still have a lot of fun building and fighting this robot.

Chapter 20 shows you how I built a competition-legal, 220-pound robot. The Six Million Dollar Mouse was built in two stages for two different television shows. The bot also competed locally until I retired it and used some of its parts to create a new, more powerful bot called MiniRip.

20

PROJECT TWO: SIX MILLION DOLLAR MOUSE

Believe it or not, the most frequently asked question about this robot is "Did you really spend six million dollars on this?" For everyone's information, no, I did not. The name came about when we finished modifying a robot frame. The frame sort of resembled a mouse. My brother and I jokingly brought up the old TV show "The Six Million Dollar Man." If you aren't old enough to remember that, the man had bionic legs, an arm, and an eye. The show aired from 1974 to 1978 which, toward the end of the run, is right about the time I started getting into electronics and robotics. The show's catch phrase was…

> We can rebuild him. We have the technology. We have the capability to make the world's first Bionic man. Steve Austin will be that man. Better than he was before. Better … stronger … faster.

It seemed really fitting, since we were rebuilding one of our previous robots to make it "Better … stronger … faster."

HandsOFF!

I'll start from the beginning instead of skipping to the upgrade. The first version of this robot was called "HandsOFF!" We were competitors in the first series of TLC's Robotica. HandsOFF! was to be a 220-pound machine with spinning saw blades in the front for show and a spinning piece of steel on the top for causing damage. It was driven with four elec-

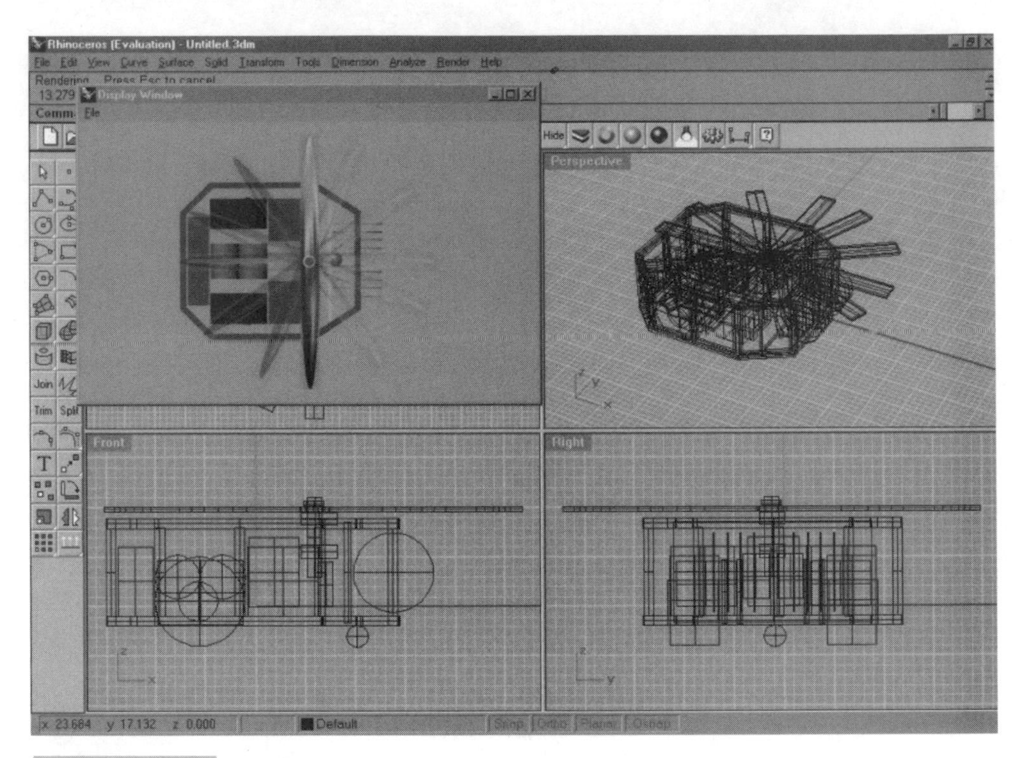

FIGURE 20.1 HandsOFF! Rhino 3D screen shot.

tric bicycle motors mounted on two custom-made gear boxes. The spinner blade was driven by a surplus military motor I picked up somewhere. Figure 20.1 shows the original Rhino 3D rendering I did.

Up until this robot, we had used sprockets and chains to drive our machines, but I wanted to do something different. I needed something to make it interesting to the TV show and something that I had not done yet. We decided to build our own gear boxes. They were highly custom, in that each box was a two-stage reduction that drove a single wheel with two motors. I broke out my Martin Gear manufacturer's catalog and started drawing up the specs for the boxes. Figure 20.2 shows what I finally came up with.

I took my gear specifications to my local industrial supply store and gave them the part numbers I wanted. I had the gears by the end of the week. I don't have space to go into the exact construction details, but after about 60 hours of work and $400 in parts I had a pair of custom gearboxes, shown in Figure 20.3, with a 16:1 reduction ratio that used two motors each.

Now I needed a frame and base plate. I had already designed the frame in Rhino 3D, and that design included measurements for as many parts as I had available. I wasn't completely sure that everything was going to fit in the space I had allowed. We were going to have to make a custom bracket to mount the spinner blade motor. As shown in Figure 20.4, we used chalk to draw the design out on the shop floor and began placing components. Once everything was in and seemed to fit, we started cutting the square tubing for the frame.

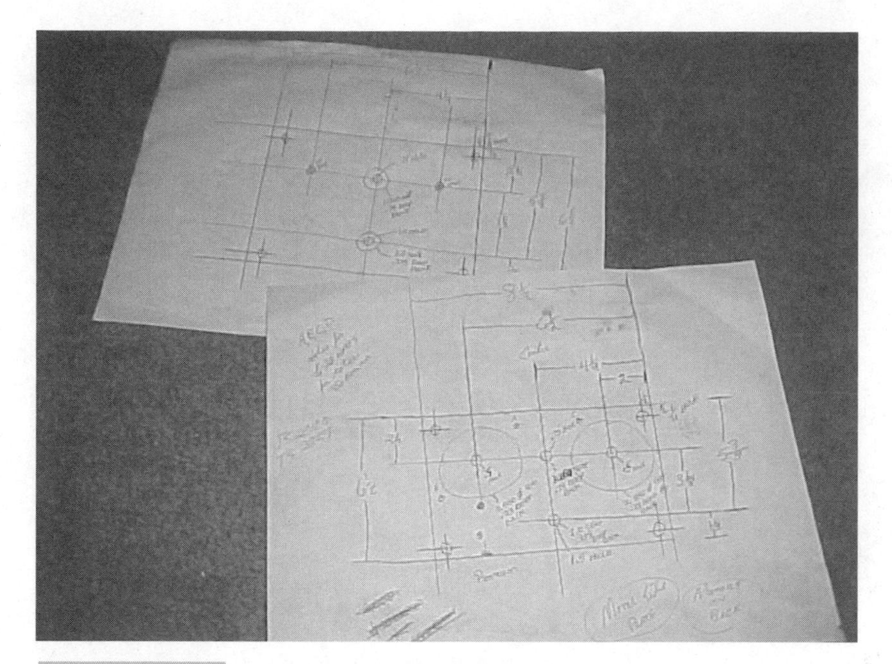

FIGURE 20.2 Drawings of gearboxes.

FIGURE 20.3 Completed gear boxes.

FIGURE 20.4 **HandsOFF! parts floor placement.**

Next up, we welded the frame together using a jig. In this case, the jig was a pattern made of several pieces of wood that had been fastened together so that we could lay our square tubing down and hold it there for welding. I'll give you a hint. Do not use wood to hold a steel frame in place. When the frame moves because of the welding heat, the wood will not do its job. Once the frame was welded together (Figure 20.5), we acquired a piece of 1/4-inch thick aluminum tread plate (Figure 20.6) and cut it to fit the frame outline with both a plasma cutter and a jig saw.

After drilling all the holes to mount the gearboxes to the base plate and to mount the base plate to the frame, it was time to mount the batteries and the spinner blade motor. We used some steel strapping material, with holes down the length, to bend and form into battery holders. The batteries were placed directly in front of the wheels to give the robot a close center of gravity. They were "strapped" to the frame sides. We placed 1/2-inch polycarbonate guards between the outside walls of the robot and the sides of the batteries for extra protection.

Now that the batteries and drivetrain were mounted, we turned our attention to the weapon systems. This weapon was a 3-inch wide piece of 3/8-inch thick steel tread plate that would be spinning around at about 900 rpm. Lots of potential for destruction! This thing was going to hit hard. We needed a strong way to mount it. We decided that a 1.5-inch diameter solid shaft would be able to handle the forces involved. Next, we needed bearings that could do the same. Fortunately, the ball bearings that are built to accommodate the 1.5-inch diameter shaft are really strong. In Figure 20.7, you see the gears used to turn the blade, the bearings, and the blade itself before we sanded it down and gave it a coat of paint.

FIGURE 20.5 Welding the frame.

FIGURE 20.6 Aluminum tread plate base.

FIGURE 20.7 Spinner parts collection.

By the time we got everything installed and almost completely wired up, the filming crew from TLC came to do our video biography. These guys were really cool, and we had a great time showing off the robot all day. One of the last things we did that day was weigh the robot. It was 34 pounds over weight, without any armor. That was bad, since we were going to use 16 AWG stainless steel. We needed to lose about 60 pounds to make weight at safety inspection. We started looking everywhere on the robot to lose weight. The first thing to go was half of the front end and all those cool looking but useless saw blades. Next we hollowed out the spinner blade shaft and drive axles. Then we dropped one of the batteries that powered the spinner and drilled lightening holes in the gear boxes. We even drilled holes in the gears and put them on a lathe to make their web a bit thinner (see Figure 20.8).

Now that the weight problem was back on track, we started mounting the armor. We gave it a black and white paint job and added a flashing blue police light to accent the "police helicopter" theme we had decided on. Figure 20.9 shows the completed HandsOFF! The competition was really fun. We met a lot of great people and saw a lot of really tough robots. We also found one of our most serious design flaws, but not until it had lost us our chance at winning anything: when we built the gear boxes, we used pins to hold the gears on the shafts. Two pins in the left gear box fell out during the first challenge. We did not know it until we had driving problems in the second challenge. Even then, we only noticed that one of them was missing. We fixed that before the third and final challenge. However, the second pin was still a problem and caused us plenty of shame and embarrassment. Nonetheless, we had a good time on our first trip to Hollywood and our first TV appearance.

FIGURE 20.8 Lighter gear boxes.

FIGURE 20.9 HandsOFF! finished.

Bot Conversion Time

We learned a lot from our experiences and were ready to put our lessons to use. The first series of Robot Wars Extreme Warriors put out a notice that they were looking for several teams to come to London, England with their robots to be in the show. We jumped at the chance. We planned to heavily modify HandsOff and rename it. I felt the modifications were so severe that a new name was in order.

The least of the modifications came about because we broke four antennas at the Robotica filming. So, we encased our antenna in polycarbonate. (You can see our mounting scheme in Figure 2.5 of Chapter 2.) The next modification was done to ensure that we would never lose another roll pin in the drivetrain. We used a broaching set and some pre-keyed shafting to convert all the gears and shafts in the gearboxes to use keys (covered in Chapter 7). The third, and major, modification was to the weapon. When I first designed HandsOff, I believed I saw a rule stating that every robot had to be at least 15 inches tall. Because of that, the spinner blade was 13 inches or so off the ground. It was just easier to build the robot that way in the eight weeks we were given. As the competition grew closer and the rules were modified, I guess that 15-inch tall rule was dropped. Because of that, we would never have been able to hit most of our opponents if we did well enough to get to the Fight to the Finish. For Robot Wars, I knew the competition would have serious weapons and we were in for some serious bot-on-bot contact. So, the third modification was to enable the blade to hit anything 4 inches or taller. We also changed the spinner blade motor to a beefy Pacific Scientific 1hp motor. That really made it dangerous. I drew up the modification plan in Rhino 3D and we started hacking the frame. We ended up with what you see in Figure 20.10. The frame with the "ears" sticking up and the spinner blade "whiskers" reminded us of a mouse.

There were the fourth and fifth modifications that really didn't affect too much. Fourth was the addition of a self-righting mouse tail. We had a dangerous weapon but we were vulnerable to low riding robots (our second major design flaw). We used an electric linear actuator and a piece of 1 × 2-inch square steel tubing to make the tail. The idea was that if we got turned over, the actuator would push the tail up and turn us over. All the tests were successful but painfully slow. We would find out later that if you can not get turned over quickly, you might as well throw in the towel. The fifth modification was to strip the black and white paint and make this thing look like the bionic mouse shown in Figure 20.11. The Robot Wars Extreme Warriors show really likes theme robots anyway. We added fur. On the sides we made it look like the fur had been ripped back, exposing the bionic insides of the Mouse. We painted the blade black with pink tips so you could see the blade when it was spinning. We added a small Nerf basketball painted pink like a nose. I created all the graphics using Adobe's PhotoShop™ software. I gave it some sinister teeth and robot-type eyes, and we were ready for London.

Competition Time

We did well in London and had a great time on top of it. The Mouse, while it did take a serious beating, never gave up and was working perfectly when we crated it up for the trip

FIGURE 20.10 Modified frame.

FIGURE 20.11 Six Million Dollar Mouse.

home. A month or so after we got back, it was time for the first NC Robot StreetFight. Two superheavy robots were signed up and one heavyweight. The heavyweight was the Mouse. I decided to put the Mouse in the superheavy category and face robots that were up to 120 pounds heavier. The Mouse survived, even though it didn't win any fights. After that, we decided to retire it and use the gear boxes and motors in our next creation called MiniRip. MiniRip is a four-wheel-drive, bulldozer-type robot with a small spinning disk on the opposite end of the dozer blade. We took MiniRip to Hollywood for the filming of series three of TLC's Robotica. It is supposed to be on TV in the summer of 2002. We feel we did well and redeemed ourselves from our first showing at Robotica. Those gear boxes haven't given us a single problem since we changed from roll pins to keys and they've been put through thirteen encounters and lots of driving practice.

Cost Rundown

That's the story of the Six Million Dollar Mouse. The only thing left are the costs. Let me run them down for you real quick:

15 Feet of 1-inch square tubing	$11.00
1 3 × 3 foot by 1/4-inch aluminum base plate (discounted through a sponsor)	$25.00
1 4 × 8 foot sheet of 16AWG stainless steel for armor	$150.00
25 Stainless steel weld nuts	$22.00
25 Steel weld nuts	$6.00
1 3 × 3 foot sheet of 1/2-inch thick polycarbonate for armor	$200.00
4 EV Warrior electric bicycle motors	$60.00
2 Custom gear boxes (gears, bearings, pre-keyed shaft, housing materials)	$450.00
2 10 × 5 go-kart racing slick tires and hubs	$40.00
2 Wheel hub adapters	$32.00
2 Caster wheels (front)	$9.00
3 Odyssey PC545 batteries	$400.00
1 3-Foot stick of heavy muffler strap	$7.00
2 4QD Pro-120 speed controllers	$380.00
1 Cast aluminum speed controller housing	$69.00
2 Team Delta 4QD interface boards	$66.00
3 Feet of 4AWG weapon motor cable	$FREE
1 Pacific Scientific 1hp motor for the weapon (huge discount at surplus dealer)	$50.00

1	Set of 4.5:1 gears for the weapon	$75.00
1	300-amp contactor for weapon motor	$40.00
1	Team Delta electronic switch	$25.00
1	1.5-foot piece of 1.5-inch diameter steel shaft (sponsor)	$FREE
2	1.5-inch ID ball bearings with four bolt flange mounts (sponsor small discount)	$70.00
1	1.5-inch ID shaft collar	$12.00
2	1.5-inch ID 3 inch OD custom spinner blade washers (sponsor)	$FREE
1	2-inch long, 1-inch inside diameter Oil-Lite style bushing	$2.00
1	500-amp main power switch	$20.00
2	150-amp emergency cutoff loops	$68.00
1	700-lb thrust electric linear actuator and necessary steel (tail components)	$60.00
1	Team Delta electric linear actuator interface	$32.00
1	Hitec Prism 7-channel radio set with spectra module	$412.00
2	Yards of fake fur	$35.00
1	25-sheet pack of photo paper	$20.00
3	Lexmark inkjet cartridges	$84.00
1	Can rubber cement	$4.00
100	Assorted nuts, bolts and washers	$19.00
10	Feet of shop vacuum hose (tail material)	$11.00
2	Small Nerf basketballs (I had a spare nose and needed it)	$5.00
Total		**$2971.00**

Since I'm sure I missed something, I feel safe calling that $3,000. None of that includes any labor costs or any tools we bought while building. It also does not include the approximate 900 man hours that my brother and I put into building and then modifying this robot.

Summary

Of the bots I've built, the Six Million Dollar Mouse was one of my favorites. It took lot of work but that is why I am involved with this sport. I love building these machines. As in Chapter 19's RAID robot, I did not show you any building steps that I took to construct the Mouse. Project two is simply here to give you a feel for the costs in time and money that a large robot project will demand.

Chapter 21 presents Project Three: Dagoth. This is a small, detailed bot and can cost between $1,500 and $2,000 to build, depending on where you buy your parts.

21

PROJECT THREE: DAGOTH

When designing Dagoth, I decided to lean toward the quickly built bot while giving suggestions on how to save some money along the way. To be truly beneficial, the motors, gearing, speed control, and radio equipment should be usable in other robots. I never liked spending money on something that would be used one time. So, for between $1,500 and $2,000, you can build a 30-pound robot and be able to use most of it in future bot projects. Only a few modifications are needed to turn this wedge bot into a 60-pound bot with an active weapon. I leave that to you, though.

The Plan

You can get started in a few different ways. The first thing I do is to formulate a plan. I usually put my plan down in the form of questions that need to be answered. Here is my plan:

1. Which competition will I enter?
2. Which weight class?
3. How will the bot move?
4. What kind of frame, armor, and weapons will it have?

This four-question plan may seem short, but it contains everything you need to build a robot. Several smaller questions subsequently get answered while answering the main four. I'll go through those in order as we discuss the plan.

WHICH COMPETITION WILL I ENTER?

You might think there is only one competition to enter, but that's just not so. As of this writing, three major television shows feature combat robots. BattleBots currently has the biggest fan base. Robot Wars, a similar English version, is a close second. Robotica, which is distinctively different from the other two shows, comes in third place. I've entered all three competitions. Each one has its pros and cons. At each competition, there is a sense of community and cooperation between the builders. Each one has its drawbacks, too. BattleBots is expensive if you do not live near San Francisco. Robot Wars only has one weight class and is entered by invitation only—however, they usually invite anyone who has a working 220-pound robot and has filled out their application. Robotica, my favorite, currently only invites twenty-four teams to participate per show. Local competitions are best. Several of these are based around the country. The oldest is the BotBash in Arizona, which usually allow four or five weight classes. You can find information about it at www.botbash.com. The next oldest is TC MechWars in Minnesota; you can find more information about this one at www.tcmechwars.com. A few competitions are held on the East Coast now. The North Carolina Robot StreetFight was the first on the East Coast to have a full-sized arena and to accept six weight classes from 12 to 340 pounds. I personally organize and produce this competition. You can get more information at my Web site at www.ncrsf.com. Since I'm building the robot in the book and I have my own competition, the answer to the first question is that we'll be entering the NC Robot StreetFight. Competition rules are very similar to the BattleBots™ rules, with the major difference being the addition of the 12- and 30-pound weight classes.

WHICH WEIGHT CLASS?

I'm a firm believer that you should stick to a smaller weight class if you are new to this sport. One reason is that a 300-pound robot costs a lot more to build in both time and money. Smaller bots are also easier to handle. Some people have jumped into building a 200- or 300-pound robot on their first try, but the results usually aren't good. My first robot was for the 200-pound class. I never finished it. For those reasons and to make sure this chapter isn't 400 pages by itself, we'll answer the weight class question by building a 30-pound machine.

HOW WILL THE BOT MOVE?

The third question, "How will it move?" is something you should think about. Several small questions are hidden inside the main question: Which motors will you use? Will it walk or roll? Which wheels will you use? Will it be two- or four-wheel drive? What speed controller will you use? You have several options at your disposal. You have to choose between a gear motor and a plain motor. If you use plain motors you will need to build gearboxes or some other form of reduction for them. There isn't much sense in building gearboxes for this robot when cordless drill motors can do the job and are so readily available.

If we check the "Motor Spec Chart Part 1" in Appendix C, we find everything we need to know about the 18 volt DeWalt cordless drill motors and gearing that I want to use. We need a total of 60 pounds of force from each motor to be able to carry our own weight plus that of the opponent. Remember that wheel radius affects the amount of torque at the floor. In this case, we're trying to find the minimum amount of torque expected of the motor so that we can determine if the DeWalt motors will supply it. So, multiply 60 pounds by the 2-inch radius of the wheel to get 120 lb-in. The Motor Spec Chart says the stall torque is 150 lb-in. That means we'll be running the motors at 80 percent of stall when we're carrying our opponent. It's only half that when driving around normally, so I'm pretty happy.

Now, I want to know how fast the robot will move at top speed. The Motor Spec Chart says the output of the DeWalt is 450 rpm. Since we decrease our torque with the radius of the wheel, we must increase our expected speed. Remember the formula for wheel circumference?

$$C = \pi \times D$$

Our wheel's circumference is 12.56 inches. That means the wheel will cover 12.56 inches every time it turns a full revolution. Since the motors are turning 450 revolutions per minute (rpm), the wheel will cover 5,652 inches every minute. Check Appendix B for the speed conversion factors to convert inches per minute to miles per hour. Multiply that by the number of inches per minute to get 5.35 miles per hour. That is a little bit slow, but with these motors we have the capability of switching into high gear, which gives us less torque and more speed. I'll leave them in low gear for now because I'm building a wedge bot and I want the pushing power.

A shortcut exists for checking the viability of the Vantec RDFR23 that I would like to use. We just said that we would normally be running at 40 percent of stall, which gives us a peak current of 44 amps. That's about the most current we'll draw, and the average will be less. The Vantec has a 30 continuous amp rating and a 60 peak amp rating. The Vantec should be fine to drive these motors. IFI Victor 883 speed controllers from www.ifirobotics.com are another choice because they can be used with larger motors. I decided to save some money and recycle the Vantec speed controller that I used in a previous robot. The Victors are actually about $40 cheaper, but require either a separate ($30) mixer or a computer radio that can mix signals. Remember, mixers make your bot easier to drive, and the Vantecs all have the function built in.

I bought the DeWalt brand gear motors from Team Delta at www.teamdelta.com, because Team Delta has created special motor mounting hardware that can be bought there as well. They also have special drive shafts, matching bearings, wheels, and hubs. On larger robots you definitely want to look around for suitable drivetrain components. The wheels are 4-inch diameter Colson brand casters, along with the special hubs they require, also available from Team Delta. Figure 21.1 shows the DeWalt Motors, Team Delta parts, and the basic parts placement.

The batteries that I want to use are from DeWalt—18-volt battery packs that normally come with a cordless drill. We need to check the numbers to make sure these batteries will last for an entire match. DeWalt batteries are of good quality, and actual testing has shown that these batteries will supply about 3 amps for an hour. Using the information and calculations presented in Chapter 5, we can figure the batteries might supply about 36 amps for 5 minutes. Remember that we calculated a maximum use of about 40 percent of the

FIGURE 21.1 DeWalt motors and Team Delta parts.

stall current, which is about 44 amps. You should also remember that we wouldn't be using that much current constantly.

The Motor Spec Chart in the appendix also says the stall current is about 110 amps, with a Kt of 21.8 oz-in per amp. Because we have a 30-pound bot and about a .9 coefficient of friction at best, the motors only need to put out about 27 pounds of force before the wheels start to spin. You can increase that number a little to compensate for friction within the chains and bearings, but we'll just keep it at 27. Convert the Kt value of 21.8 oz-in per amp to 0.114 lb-in per amp. Now, figure how many amps will be drawn when the wheels begin to spin. If the motors are putting out 27 lb-in at 0.114 lb-in per amp, then we must be drawing about 3 amps. Don't forget to double that amount because we designed the bot to be able to carry its opponent. Just by the nature of the batteries, we know that we can draw about 3 amps for an hour. However, we will draw more amps than that at several points during the match. Every time we start moving, the motors draw more current. Every time we try to turn or reverse direction, the motors draw more current. We'll consider our minimum current draw, when moving, as 3 amps. So, I feel pretty safe using a single battery pack to power this version of the robot. Later on, we'll test the robot and make sure that the batteries will last the entire match.

WHAT KIND OF FRAME, ARMOR, AND WEAPONS WILL IT HAVE?

There are some really tough 60-pound robots in the world today. I personally plan to compete with the robot I build in this book in that weight class after I add an active weapon. (I won't show how I added the weapon; in part because of time and space constraints, but more important because I think builders should use their own creativity in the design of their robot. I do not want to give you blueprints so that you can buy the parts and have someone put it together for you. If you're reading this at all, I think you'll agree with that.)

I'll start the frame with a base plate made of 1/8-inch thick, 6061 aluminum. The robot will be 4 inches tall only because the wheels are that tall. The body of the bot is actually 3 1/8-inches tall, so that the top and bottom of the wheels protrude to make the bot invertable. The side walls will be 1/2-inch thick polycarbonate. The front and back will be 1/2-inch thick, 6061 aluminum. The top will be 1/4-inch thick polycarbonate. The Rhino 3D rendering in Figure 21.2 shows what I have in mind.

In this case, the frame is the armor. Later, we'll add some gussets and other supports to make it stronger. Dagoth will have two weapons in this version, both inactive but effective. Three, 4-inch, hardened steel spikes will be mounted on the front. These will hopefully be used to puncture tires and light armor. They also serve as armor when you confront a spinner bot. These spikes will take the beating and may slow the spinner down enough for you to attack with the wedge. The wedge can be made from any number of materials. My plan is to construct it from 16 AWG stainless steel, with the possibility of replacing it with 16 AWG 6-4 titanium. That might be an expensive move, but I think it will be worth it. Figure 21.3 shows the entire design in a Rhino rendering, minus the top and the wedge for clarity.

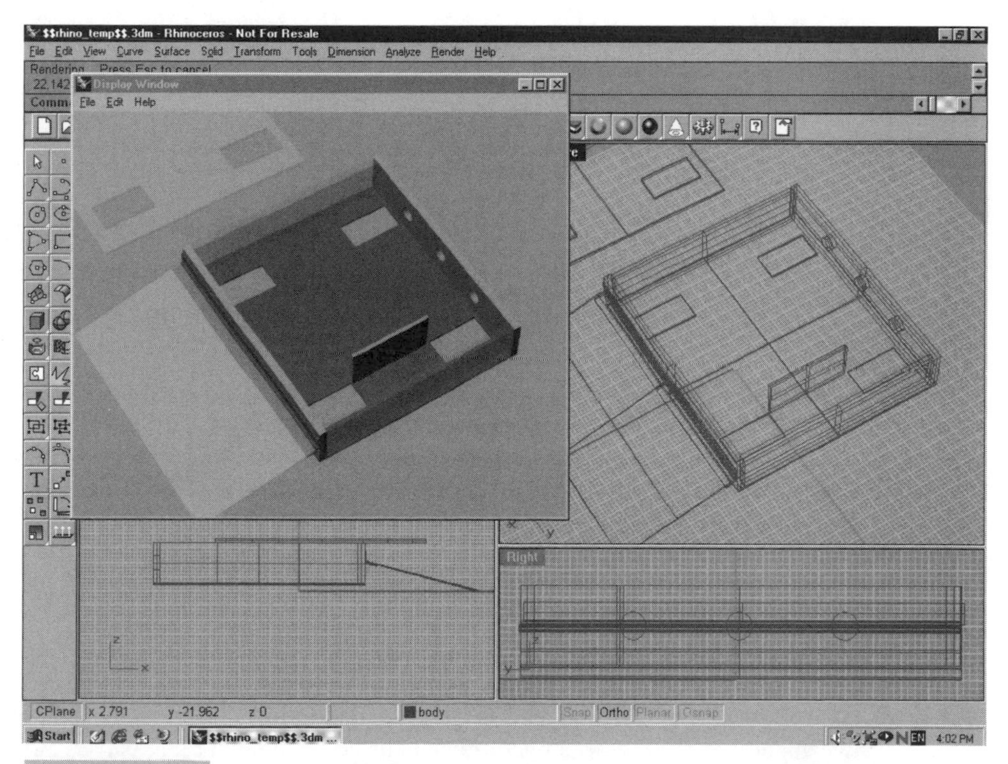

FIGURE 21.2 Rhino render of body only.

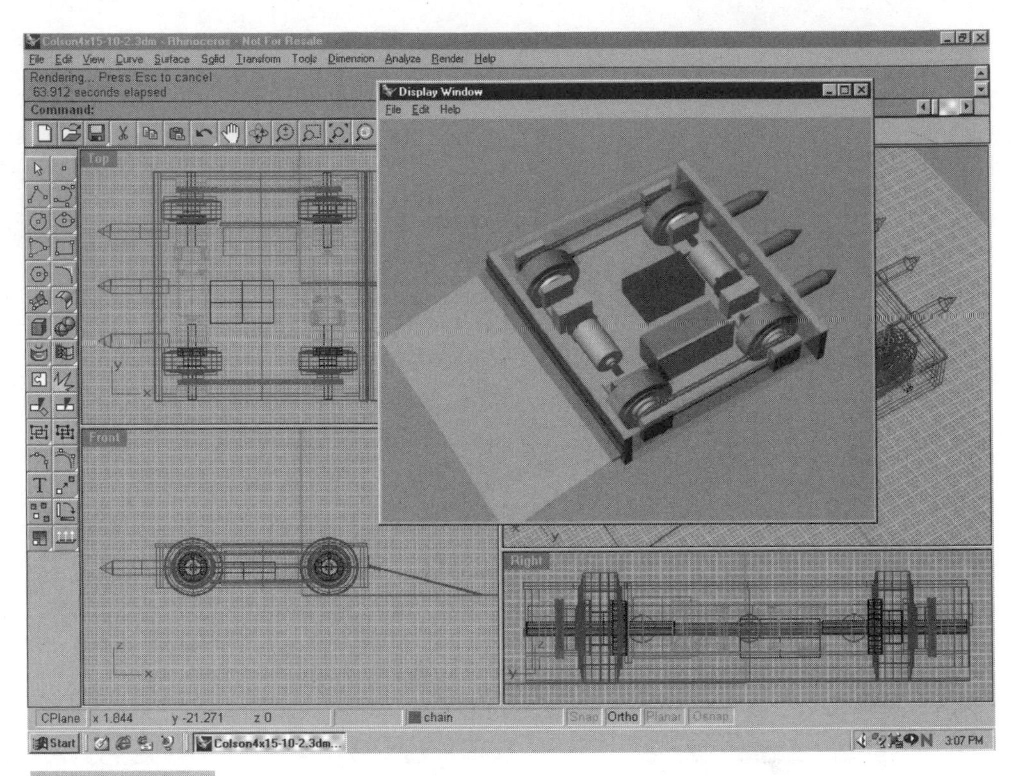

FIGURE 21.3 **Rhino render of Dagoth.**

Let's go over our plan again, with answers this time:

1. Which competition will I enter?
   NC Robot StreetFight at www.ncrsf.com.
2. Which weight class?
   30-pound lightweight.
3. How will the bot move?
   Four wheels. Two-wheel drive using 18-volt, DeWalt cordless drill motors. Vantec speed controller. 18 volt, 2ah DeWalt cordless drill batteries.
4. What kind of frame, armor, and weapon will it have?
   6061 aluminum base, front, back, and internal supports. Polycarbonate sides and top. Stainless steel wedge and hardened spikes.

Mechanical Design

The mechanical design of this robot is fairly simple. In fact, I'd say there is nothing innovative about it at all. The armor and frame boils down to eight main parts. First, the base plate is made of 1/8-inch, 6061 aluminum, and it will hold all the "guts" of the robot together. The motors, four of the eight bearings, two battery packs, speed controller, the radio receiver and other assorted electronics will be mounted directly to it. The internal supports and side, front, and rear armor will also be attached to the base plate.

The overall design is an aluminum and polycarbonate box. Three sizes of bolts are used in Dagoth. I'm using several quarter-twenty thread by 1-inch lengths and several 3/4-inch lengths. The DeWalt motor mounts use 10-24 thread bolts in 1-inch lengths. All of the gussets use the same size bolt. Each bolt is the countersunk version of the socket-head cap screw.

The Team Delta DeWalt motor mounts do not support the motor itself. Because of this, the motor can break free of the transmission housing's plastic mounting holes. The DeWalt motors should be supported on all sides. My special mount does not support the motor on the top but it does cover both sides and the bottom. Figure 21.4 shows a Rhino rendering of the special motor supports. Note that these are bolted to the base plate and aluminum armor with 10-24 threaded countersunk bolts. Each motor has its own support bracket. The bracket itself is also performing a gusset function, which makes the frame stronger.

Several small, triangular gussets are positioned around the joint between the armor and the base plate. Gussets are also necessary around the joint between the armor and the polycarbonate lid.

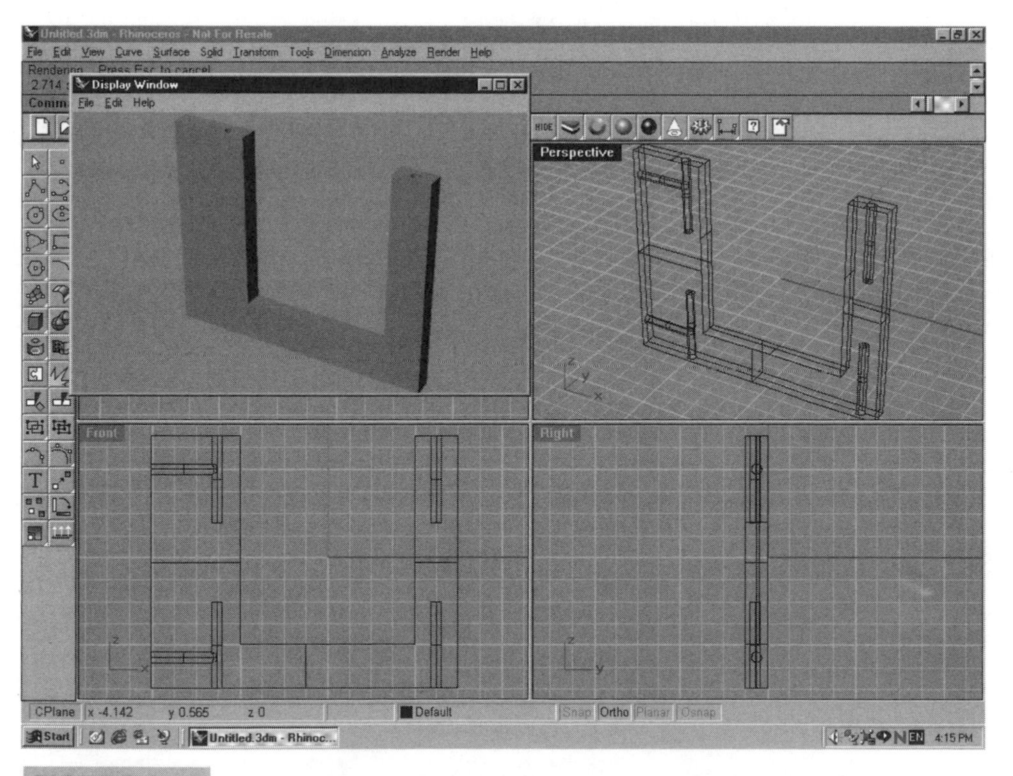

FIGURE 21.4 **Rhino render of special DeWalt motor support.**

Where the Parts Go

Since I have access to a machine shop, I saved a little money by doing the entire cutting and finishing of the body myself. This included using the plasma cutter to cut the aluminum and using the milling machine to get square edges and to get the parts to finished size. All aluminum is of the 6061 variety.

The base plate is 16 × 14 1/2-inches and is 1/8-inch thick. This may seem a little thin, and it would be if we weren't planning on adding some extra support. I do plan to add some 16 AWG titanium sheet or spring steel sheet to the bottom for both support and protection. However, I will not be doing that in this version of the robot. Four large holes in the base plate accommodate the wheels. The holes are 3 1/2-inches long by 1 7/8-inch wide. This is not a totally arbitrary hole size. As you can see, I used Rhino 3D to put the design on the computer. Once I had the wheels placed on the base plate, I simply moved the base plate up a bit from its planned position. Then I used the software's Boolean Difference function to cut holes in the solid plate based on the size of the wheels. Next, I measured the dimensions of the newly created holes using the Length function. I then created a slightly larger box. This box was larger to account for any errors on my part. Once again, I used the Boolean Difference function to cut new, larger holes for the wheels using the larger box. The hole does not need to be the full size of the wheel; I only want 1/2-inch of ground clearance under the robot, and this arrangement gave it to me.

I used a plasma cutter to cut the base plate from the stock material, filed down the edges using a coarse file, and then finished the edges using an electric sanding machine. I did not immediately cut the wheel holes into the base plate, although I wished I had. Instead, I put that part off until later and used a hand drill and a jig saw to make them. I used a file

FIGURE 21.5 Cutting wheel holes in the base plate.

to clean up the edges and then a bit of sandpaper to polish them up. Figure 21.5 shows the base plate in the middle of having the wheel holes cut out.

The top armor was designed and cut in a similar manner on the computer and in the shop. The top armor is made of 1/4-inch thick polycarbonate. I did not use the plasma cutter to cut the polycarbonate; I used the band saw. Large teeth and a quick blade speed will do wonders when cutting most plastics. (Take a few practice cuts on some scrap before you cut a good piece of polycarbonate.) I used a jigsaw to cut the wheel holes. The top armor dimensions are 16 × 14 1/2-inches.

The front and rear walls of the bot are 1/2-inch, 6061 aluminum. When this bot slams into its opponent with the front spikes, all the force is concentrated into two or three small spaces of the front wall. The wall needs to be able to handle the load. Because of the main internal supports, the force experienced by the front wall is dissipated into the entire frame rather than into the front wall alone. The rear wall is armed with a wedge instead of spikes. The wedge spans the length of the wall and spreads the force load out across more of the wall. The main internal supports also help with distributing force loads from the rear wall into the rest of the frame.

MOTORS

I am using DeWalt brand, 18-volt, cordless drill motors, and the custom motor mounts available from Team Delta. There are a few things you need to do in order to use these motor mounts, which come in three parts. Team Delta includes an instruction sheet, but I figured I would go over it in a little more detail. Figure 21.6 shows everything needed to use these motors and mounting brackets.

FIGURE 21.6 DeWalt motors, bracket, and tools.

Very little work is necessary on the motor itself. The motor face is a mounting bracket that holds it to the gearbox. Four holes on the bracket must be tapped with a 6-32 thread. (Tapping is the act of cutting threads into a hole so that you can use bolt to hold two things together. Check out the Tools section of the book to find out how to tap a hole.) When I tapped the four holes on my motors, the material felt grainy and I had no problem cutting the threads without tapping fluid. The main problem I see is that you must run the tap to within 1/4 inch of the neck of the tap to get full threads in the material. This was not a major problem, but I did notice that the end of the tap needed to cut away a small amount of the motor casing. Once you get to this point, you should slow down a bit. The tap also tends to drop material chips as it cuts. If you hold the motor in a certain direction you will get aluminum from the motor face plate, and even steel chips from the motor housing, into the vent holes of the motor. Pay attention to this because those chips can cause shorts, or they can get between the magnets and the rotor. Neither of these prospects is favorable, so you may want to cover the vent holes with some electrical tape. Remove the tape before operating the motors. Figure 21.7 shows one of the mounting holes being tapped.

The second part of motor construction is the gearbox or transmission. This planetary gearbox is commonly found in cordless drills because of its great strength and high gear ratio possibilities. From Team Delta, the gearbox comes with a white plastic cap over one end, which is held on by a rubber band. The cap holds the spindle into the housing. If you remove the cap, it is no difficult job to tip the gearbox over and dump out all the little gears—something you do not really want to do. Figure 21.8 shows the partially disassembled DeWalt transmission.

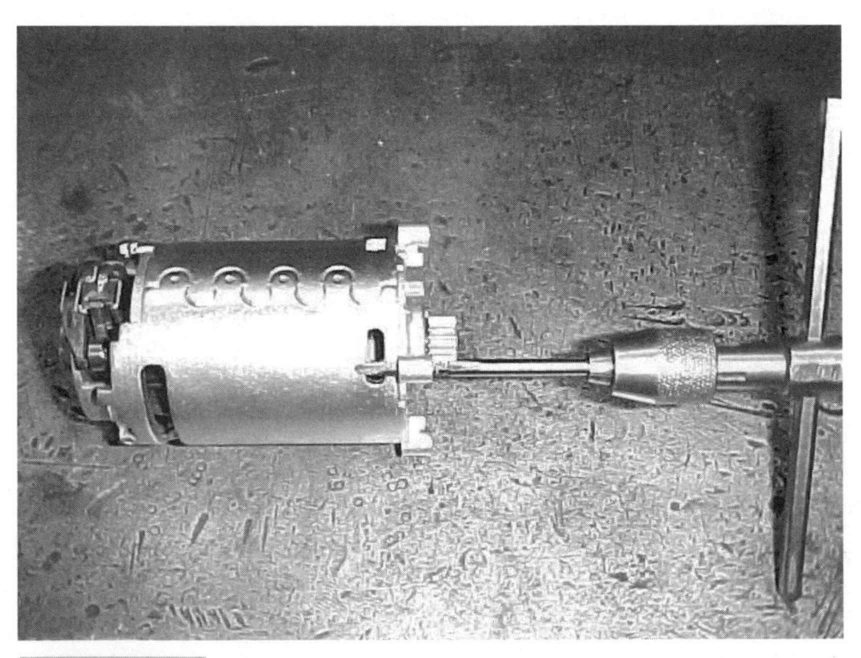

FIGURE 21.7 **Tapping the motor mounting holes.**

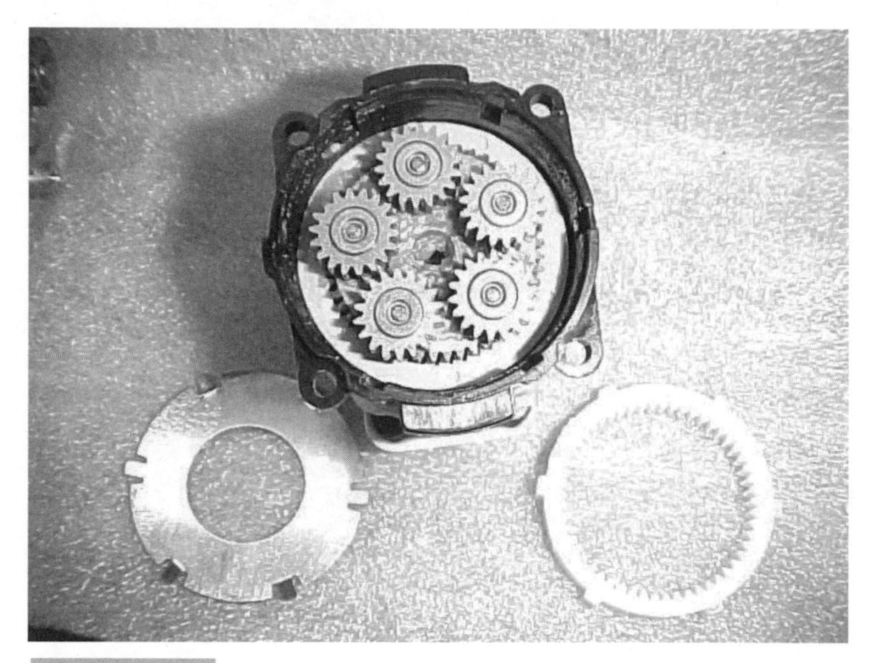

FIGURE 21.8 **Partially disassembled DeWalt transmission.**

Something you really do want to do is install the low-speed locking ring if you plan to run the motors in low-speed, high-torque mode. I am running in low-speed mode, so I need to install the ring. The arrow in Figure 21.9 shows the installed ring. The ring locks the large yellow gear in place. Use a small screwdriver or knife to remove the thin metal plate on the end of the transmission. You will see five small metal gears inside a large greenish plastic gear. Remove the plastic gear and set it to the side. Insert the low-speed locking ring, and then replace the white plastic gear. You must move the small metal gears individually to get the greenish gear installed. Next replace the thin metal plate. You may have to move all the guts of the transmission around to get the thin metal plate reinstalled.

If you decided to go with high-speed instead of low-speed operation, or you try the low speed and figure it's not for you, you must instead pull the five metal gears and their mounting piece out of the transmission. Then pull the yellow gear out. Turn the yellow gear upside down and put it back into the housing. Replace the metal gears, the greenish gear, and the thin metal plate. This will have your bot running faster, but it won't be as strong.

Use some 6-32 1/2-inch long bolts to mount the motor to the transmission. Use socket head bolts instead of the hex head bolts and install them in the threaded holes we just made. (You might have to wiggle the motor gear around to get it lined up with all the small gears in the end of the transmission but you will eventually get it. The transmission will fit the motor correctly in one orientation, so if it is not sitting flat, it is not installed correctly.)

Once you have the motor bolted to the gearbox you need to tap four more holes. Although it is not recommended, you can use self-tapping screws instead of bolts. Whichever way you plan to attach the mount to the gearbox, it must be sturdy and battleworthy. You make the decision. I spent 5 minutes tapping the holes.

FIGURE 21.9 **Low-speed locking ring installed.**

The mount itself has a flat bottom with four tapped holes. The top has a small hollowed out portion on the inside. The gearbox has an arrow molded into it. This arrow should line up with the hollowed out portion of the mount. Push the mount onto the gearbox with your hand. Once you can't push it any farther, check that the spindle is not catching on the output hole of the mount. Move it around. Then, using either a soft mallet or a hand, pop the mount completely onto the gearbox. The spindle will be flush with the motor mount when it is correctly installed. Use the four-socket head bolts included with the mount to secure the gearbox to the mount. If you overtighten these, you might break the plastic. The only thing left to do is test that the motor still turns now that you have messed around with the gearbox. I installed leads on my motors and used a battery. Follow these same steps for both motors.

If you decide not to buy the mounting hardware for the DeWalt motors from Team Delta, or if you decide to use cheaper drill motors, you must come up with a mounting scheme of your own. I sometimes buy cheap cordless drill motors at flea markets. (Actually, I buy the entire drill.) These sometimes come with two battery packs and a charger for as little as $40 for 14.4-volt to 18-volt drills. That's $40 as opposed to the $145 for just the motor, gearbox, mounting hardware, and a drive shaft from Team Delta. Since I was on a tight time schedule and I wanted an easily mounted motor solution, I skipped the cheap drills.

That wasn't the case on the last robot I built. I used cheap drill motors that were already attached to the gearbox, but still had to do some modifications so that they could be used as reliable drive motors. Most drills have a clutch mechanism that allows the motor to continue spinning, even though the drill or screwdriver bit is held still, to save battery and motor life. (The Team Delta mount takes care of this clutch modification for the DeWalt motors.) The clutch usually consists of a couple rings with several ball bearings. Depending on the position of the clutch, the rings squeeze the ball bearings onto a gear, tightening the grip of the gearbox to the spindle. The top ring usually presses on a second ring via springs. The tension on the springs dictates the slip of the gears. You must measure the length of the spring when it is fully compressed and replace them with something solid of that length. As shown in Figure 21.10, I replaced the springs with small aluminum cylinders I bought at RadioShack.

Once the clutch modifications are done, you need a way to mount the motor to the base plate. I chose to build the simple motor mounts shown in Figure 21.10 out of 1/2-inch thick polycarbonate. First, I cut the polycarbonate into four rectangles of about 4 × 2 1/2 inches. Next, I used a hole saw that was the same diameter as the motor and gearboxes. I put a hole in the center of all four pieces of polycarbonate. The next step is to drill and tap four holes in each piece. Drill parallel with the large flat side, on each side of the hole, using a #7 drill bit. Drill completely through the material. Finally, I used a bandsaw to cut the four pieces in half, across the center of the circle and long ways on the rectangle. You should have two pieces of polycarbonate with half of a large hole each. Take the top half of the part and re-drill the #7 holes with a 5/16-inch drill bit. Take the bottom half of the part and tap the two #7 holes with a quarter-twenty tap. Go all the way through if you can. If you cannot, turn the part over and start on the untapped end of the hole, going as deep as possible.

Place the motor into the bottom half of the part. Place the top half of the part in position over the motor. Use quarter-twenty bolts to secure the top half to the bottom half, firmly squeezing the motor in the process. Do not tighten too much, or you may crush the motor housing. However, the motor should not be able to rotate within the part. If it is loose because of the inaccuracy of the holesaw, use tape or a mouse pad to build up the diame-

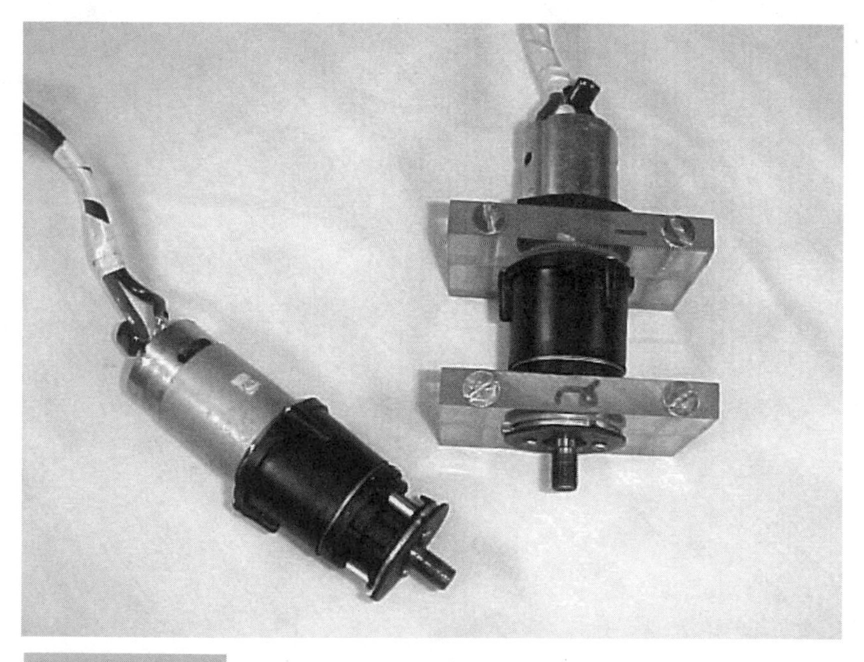

FIGURE 21.10 **Cheap drill motors.**

ter of the motor. Use two of these new mounts per motor. Be sure to get the bottoms aligned with each other so that they sit flat on the mounting surface.

MOUNTING THE WHEELS, BEARINGS, AXLES, AND MOTORS

I bought special pillow block bearings from Team Delta (shown in Figure 21.11). They are meant to support an overhung load, and although I do not intend to have an overhung load, I figured I could buy these bearings and modify them more cheaply than buying separate bearings.

First, I used a vertical band saw to cut one leg off the bearing. This left me with a flat, flange-mount bearing and an L-shaped pillow block bearing. The bandsaw is not the cleanest cutting instrument in the world—the mess is shown in Figure 21.12.

I clamped each separate piece into the vise on the milling machine. I used a 1-inch diameter end mill to face off each new bearing to the same size with a nice clean edge, as shown in Figure 21.13. A little bit of sanding made them look as if they were bought that way. The L-shaped pillow blocks retained the threaded mounting holes that were there when I bought them.

The new flange-mount bearings had no mounting holes. Using the scale and a scribe, I marked two hole centers in the corners of the aluminum block. I drilled the holes using a #7 drill bit, because I planned to tap them with a quarter-twenty tap. (Check the table of Drill Sizes and Threads in Appendix C for other drill and tap sizes.) Do not forget to use some lubrication when drilling and tapping. (Check the Tapping Lubricants Chart in

FIGURE 21.11 Team Delta bearing blocks.

FIGURE 21.12 Using the bandsaw on the bearing blocks.

FIGURE 21.13 Cleaning up with the milling machine.

Appendix C for the appropriate tapping lubrication types.) Use the same type of lubricant when drilling.

Now that the bearings are prepped, we can move on to mounting everything. If you did everything the same way I did, you will notice that the wheels do not sit freely in the holes in the base plate and match up to the motor. I designed the wheel holes so that they allow only a bit more wheel than is required to have a 1/2-inch floor clearance. Thus, the motors and bearings must be mounted above the base plate. The actual measurement from the bottom of the motor mount to the base plate is 1/2 inch. So, we need a 1/2-inch thick spacer with the same hole pattern as the motor mount and the bearing. You can get really fancy when designing these spacers, but I decided to use some scrap aluminum I had lying around. Figures 21.14 and 21.15 show the spacers I made for the motor mounts and the bearings. I cut the blocks to size and drilled the appropriate holes based on the blueprints provided with the parts. The only thing I needed to do was make sure the axle and other parts were lined up with the center of the base plate's wheel hole. Since I now had parts that had the same hole pattern as the motor mount, I could use a hole transfer to mark where the holes needed to be on the base plate. I also used the hole transfer to mark where the L-shaped bearings would go, right next to the spacers. The hole transfer is covered in the Tool section (Chapter 15). Each wheel needed spacers. The wheels without motors are mounted in the same manner.

I used 4-inch Colson caster wheels with special aluminum hubs from Team Delta. The hubs were bolted into the wheels. Each hub has a key slot so that it does not spin on the shaft, but these wheels do not have setscrews, so they need to be sandwiched between two bearings or between a bearing and a shaft collar. In this robot, there will be a sprocket next to the hub. Everything was held together with a shaft collar, and the open end of the shaft was to go in the square half of the bearing I previously modified. Figure 21.16 shows this wheel–sprocket–collar sandwich.

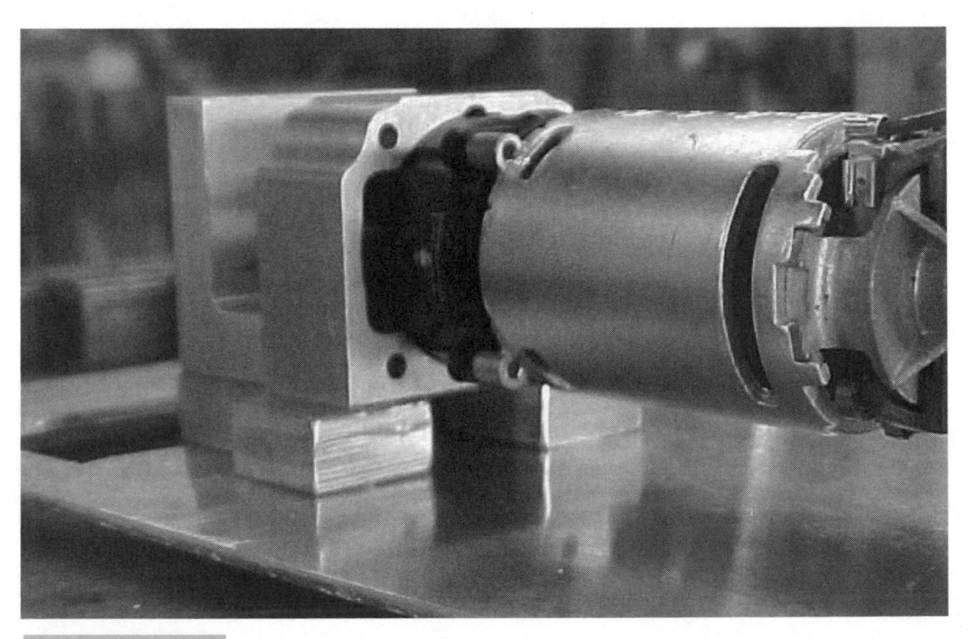

FIGURE 21.14 **Motor mount spacers.**

FIGURE 21.15 Bearing spacers.

FIGURE 21.16 Wheel, sprocket, and axle mounting.

The square part of the bearing was mounted to the 1/2-inch polycarbonate wall of the bot, as shown in Figure 21.17. After mounting the L-shaped bearing and spacers, I used a hole transfer, placed inside the bearing, to mark where the axle center would be on the polycarbonate wall. Now that there were two axle center marks on the wall, I placed the hole transfer back on the mark. Then, I placed the square bearing in place on the transfer and clamped it down. Next, I marked the two mounting holes for the square bearings. I drilled the two mounting holes for each bearing as well as the axle centers. I drilled the axle centers so that I would have a way to tap on the end of the axle, if needed, to get it into the correct position.

Before mounting the square bearings, I had to mount the polycarbonate walls to the base and front and rear panels, as shown in Figure 21.18. Using a scribe, a punch, and a scale, I marked holes along the edge of the base plate. Each hole was 1/4 inch from the edge so that it was directly in the center of the polycarbonate wall. The two end holes were placed 1 1/4 inches from the front and back edges. The other two holes were placed 4 inches in from each end hole. I used the same process to mark and place the holes that mount the front and rear aluminum armor. Two, quarter-twenty by 1-inch bolts bolted each aluminum plate through the ends to the polycarbonate plates. Now that the sides were mounted to the base plate, I could install the axles, wheels, and sprockets.

The shafts were specially made to work with the spindle of the DeWalt motor. The spindle end of the shaft was hardened; the rest of the shaft had a key slot that matched the one in the wheel hub. The shafts needed to be cut to size. The driven axles were 4 1/2 inches long; the nondriven axles were 3 1/2 inches long.

FIGURE 21.17 Square bearing mounted on polycarbonate.

FIGURE 21.18 **Polycarbonate walls mounted to base and armor.**

The sprockets I used to transmit power from the driven wheels to the nondriven wheels were from Martin Gear Inc., for a #35 chain with 15 teeth and 1/2-inch bore. The bore itself is in a bronze bushing that has been pressed into the sprocket. You can use any sprocket you like, as long as it fits. These sprockets were originally meant to be idler sprockets. (Idler sprockets spin freely on a nonkeyed shaft while keeping the chain slack down to a minimum. They can also be used to increase the number of teeth on the driven sprocket that come in contact with the chain.) Since I had these in the junk box, I decided to use the broaching set to cut a key slot into the bronze bushing. Cutting a keyway weakened the bushing and, therefore, the press fit's strength. To keep the bushing from spinning inside the sprocket, I added a setscrew that, when tightened, pressed against the key material.

After mounting the motors, bearings, axles, wheels, and sprockets, it was time to run a chain from the driven wheels to the nondriven wheels. I did not bother calculating the chain length beforehand. In the junk box, along with the spare sprockets, I had a 7-foot length of #35 roller chain. Normally, chain comes in 10-foot lengths but I had used some of this chain before this project. I also happened to have several half links and master links, which are used to tie a chain together at the ends to form a continuous loop. Half links are used when the length of chain you need is too short or too long to be attained within the length of one link. Figure 21.19 shows a half link and a master link.

Wrapping the long piece of chain tightly around both sprockets, I determined the length I needed to make the loop. Using a chain breaker is fairly simple. First, grind the head off of the roller pin you need to remove to break the chain. Open the jaws of the breaker and place a single link within them. Twist the top handle clockwise to press out the roller pin. It turned out that I definitely needed the half link. Half links should be used along with a

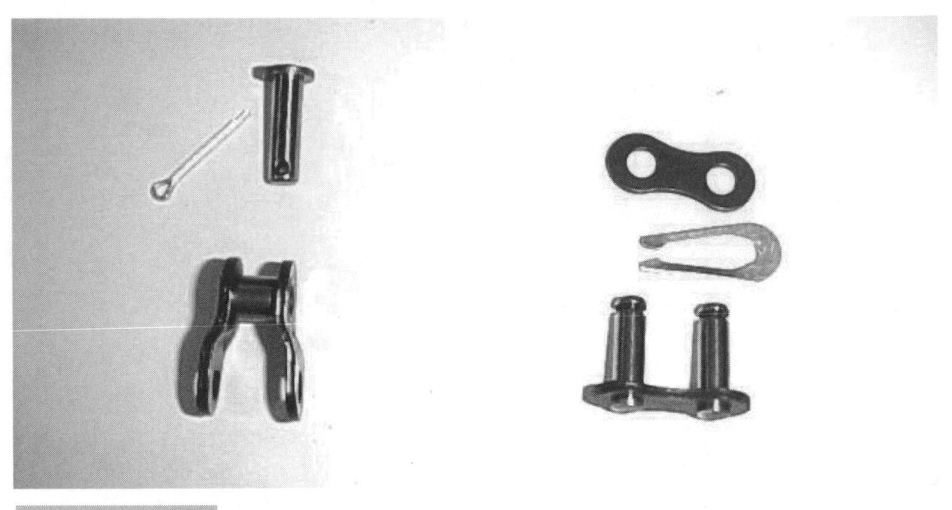

FIGURE 21.19 **Half link and master link.**

master link, so I had to remove one more link from the new short piece. Figure 21.20 shows the half link and the master link in use.

There was still quite a bit of slack in the chain, as shown in Figure 21.21, once it was put together. I needed to add a chain tensioner to tighten it up.

I had a small block of Teflon sitting in the junk box, too. By using my finger to pull the bottom of the chain tight, I measured the distance from the bottom chain to the base plate of the bot. Then, I cut some rectangles from the Teflon to the height required, about 1.5 × 1 × 1/2 inches overall. On 1/2-inch grinding wheel, I ground a shallow guiding groove in

FIGURE 21.20 **Half link and master link in use.**

FIGURE 21.21 Slack chain installed.

FIGURE 21.22 Teflon blocks mounted.

the tops of the Teflon blocks. On the bottom, I drilled and tapped two holes for mounting to the base plate. Mounting the Teflon blocks directly beneath the bottom of the chain (as shown in Figure 21.22) was the perfect thing to take out the slack.

Electrical Layout

Since Dagoth does not have any active weapons yet, the electrical side of the bot is fairly simple. There are five major electrical component categories here.

- Motors
- Batteries
- Receiver
- Speed Controller
- Wiring

MOTORS

Now that we've covered the mechanical, mounting side of the motors, only two other things need to be done to them. The first and most important thing to do is add the RFI suppression capacitors. In Chapter 2, I covered RFI suppression capacitors and I offered a schematic on how to wire them up. In this case, I'm only going to use one capacitor per motor, as shown in Figures 21.23 and 21.24. I'm using a 0.1 μf (microfarad), 75-volt capacitor from RadioShack.

I soldered the capacitor leads to the crimped end of the brush housing. Then, I insulated the connections and the entire brush housing plate with Grip-Guard®, as shown in Figure 21.25. Grip-Guard® is a liquid electrical tape or liquid rubber and is shown in Figure 21.26. Once dry, it is an electrical insulator like the rubber grips on the handles of metal tools. Apply it generously to completely cover the contacts as well as the leads of the capacitors. This will keep the leads from shorting to the motor case and will guard against possible shorts from any loose metal flying around inside your bot.

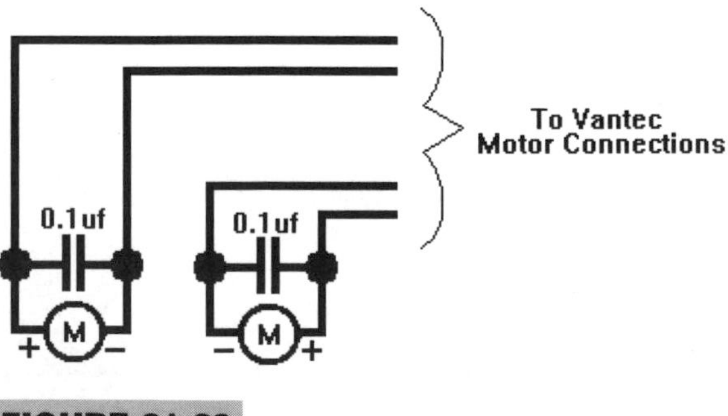

FIGURE 21.23 Motor capacitor hookup.

FIGURE 21.24 Capacitor solder connections.

FIGURE 21.25 Capacitor treated with Grip-Guard.

FIGURE 21.26 Can of Grip Guard®.

BATTERIES

Two sets of batteries reside inside Dagoth. The first and main set is an 18-volt DeWalt battery pack that I bought from a hardware store while at a competition in Florida. The second set is the four-cell, NiCd pack that powers the remote control receiver.

MAIN DRIVE BATTERIES

If you have ever seen a cordless drill, you know the batteries come in a plastic form that usually fits in the drill handle. I removed the plastic housing to make the pack lighter and smaller. I clipped the charging tabs from the pack and soldered 10 AWG wire in their place. This wire is used to connect the pack to the rest of the bot. I added Anderson Power Pole brand 45-amp connectors to the ends of the wires to make it easier to switch out charged and depleted packs. Once the pack was bot-ready, with the appropriate wires and connectors, I wrapped it with electrical tape to make sure no cells would short out on the base plate or because of flying metal debris.

I mounted the battery pack to the base with Velcro straps and wire ties. First, I made a battery support plate from 1/4-inch thick aluminum. In the plate, I cut four slots with a milling machine. I used a file to smooth the edges of the slots so that they would not cut the Velcro straps. The slots, and therefore the straps, were positioned so that the battery would be supported on two ends. The plate was drilled and tapped on one edge, so that it would mount vertically to the base plate. I added some double-sided tape to the bottom of the battery just to give it a little more sticking power. Figure 21.27 shows the drive battery mount setup.

Receiver batteries. The second battery pack is a four-cell, 700-mAh, NiCd pack meant to power the remote control receiver. This pack is available at any hobby store that

FIGURE 21.27 Drive battery mount.

carries remote control airplane parts. This small battery pack was mounted to the base plate using double-sided tape and wire ties. The wire ties penetrate the base plate. Because of the light weight of these battery packs, wire ties are usually sufficient to hold them down. If you want to be ultra safe, use the same mounting method as I used on the drive battery pack.

Charging the batteries. I do not own the Astroflight NiCd battery charger that I mentioned in Chapter 5. When I bought the DeWalt battery pack, I also bought a DeWalt 18-volt charger. Since I removed the charging tabs from the battery and replaced them with wires and a connector, I needed to do the same thing to the battery charger. With the charger unplugged, I removed the cover and followed the charging leads to the circuit board. Next, I snipped those leads and replaced them with 10 AWG wire and the mating connector for the batteries. Then, I pushed the connector through the charging clip hole and screwed the cover back down onto the base. After that, I had an 18-volt charger for this and other battery packs even though it was slow. The charger and the batteries were both floor models that the hardware store had on a shelf in the stockroom. They could not sell them as new, so they gave me a great price. All I had to do was ask if they had anything like what I needed.

Along with the receiver battery pack, I used a switch and charging harness. This switch has two functions. When turned on, it allows power to the receiver. When turned off, the batteries are directly switched to a female connector to which the charger can be engaged (see Figure 21.28 for the schematic). The switch and charging harness can be purchased at the same hobby story that sold the battery pack. If you bought your remote control gear

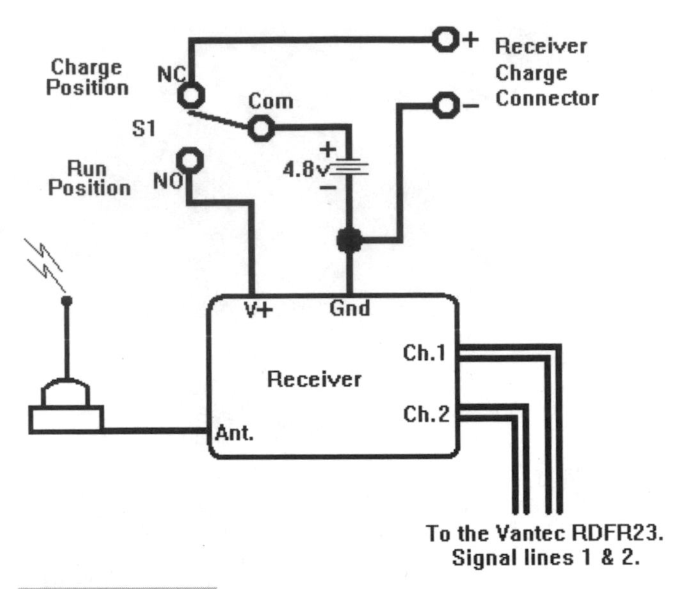

FIGURE 21.28 Schematic of receiver battery switch.

new, the package probably included at least one battery pack, the switch–charging harness and the charger.

I replaced the hobby switch with a larger, sturdier version that could be mounted on a small block of polycarbonate. The block was then drilled, tapped, and mounted to the base plate of the bot. I used a single, self-tapping screw to mount the switch to the block. When you try this, drill a pilot hole for the self-tapping screw. If you do not, the polycarbonate will gum up and stick to the screw, thus destroying the threads that you are trying to create. It will also be next to impossible to get the screw completely into the material, and the screw will probably break off inside the block.

RECEIVER

Three connections to the receiver must be dealt with. We just talked about the receiver battery connection, and we will talk about the receiver connections to the speed controller in the next section. The antenna connection is described here. As discussed in Chapter 2, the antenna on a new receiver is simply a long wire, which can get in the way of the internals of the bot. Because Dagoth is an invertable bot, we can hardly expect to be able to run the wire straight out the top. It will inevitably get snagged on a wheel and ripped off if the bot gets flipped over. Optimum placement of the antenna in this type of radio system is straight up and down.

I did not want to deal with the long wire, so I installed Dean's base-loaded whip antenna. (Directions for doing so are in Chapter 2.) You may want to experiment with different mounting methods that place the antenna horizontally inside the bot. I used a thin piece of plastic as a mount. I used a small propane torch to heat it and bend it into an L shape. The bottom of the L was mounted on the base plate, and the top of the L extended up to the polycarbonate lid. The antenna was mounted on the top of the L as close to the lid as possible, as shown in Figure 21.29.

FIGURE 21.29 **Antenna mounting bracket.**

Mounting the antenna so that it sticks out the top would give us the best possible reception, but it might also inadvertently cause the worse possible interference if the bot gets flipped. While flipped, the antenna is scrubbing on the floor. Most places where you will be driving the robot will have a rough surface for traction, and the roughness will wear the insulation off the antenna. The floor will probably be metal, too. Contact between the metal floor and the bare antenna wire will introduce a lot of noise to the receiver, maybe enough to cause it to shutdown. The best thing to do is experiment and find out which way works best for you.

Mounting the receiver module itself was done in the same manner as the receiver battery pack. Four small holes were drilled in the base plate. Instead of double-sided tape, I used more mouse pad material to give the receiver a little shock protection. I used hot glue to hold the mouse pad to the receiver and used small wire ties to attach it all to the base plate.

Receiver connections are usually a snug fit, but sometimes our robots experience very large shocks in the middle of battle. These impacts can cause connectors to jump out of their socket. This will kill your bot faster than anything. I used hot glue to secure the battery and speed controller connections to the receiver. My receiver had a cover over its crystal, and I simply put a piece of tape over the entire cover to hold it in place.

SPEED CONTROLLER

I used a Vantec RDFR23 for speed control. The larger Vantecs have a screw terminal strip meant to be used with standard ring connectors. The smaller Vantecs, such as this one, have a large housing for all the connections. Wires are inserted at the side and a small screw on the top is used to press down on the wire and secure it inside the hole between contacts. The Vantec, as well as other speed controllers, should be mounted so that it is not in direct contact with the frame of the bot. This helps protect the unit from impact shocks.

Mounting. In this case, I used the trusty old mouse pad method of mounting. Because it would not be pulling the maximum amount of current, extra heat sinking was not necessary. As insulation from metal debris, I wrapped the Vantec in black insulating tape, covering the large opening in the case. I left the connector housing exposed. Next, I cut three pieces of mouse pad. One piece was cut to fit the entire flat of the back of the Vantec. The next two pieces were cut about 1 inch wide and 2 inches long. Figure 21.30 shows the Vantec, hose clamp, and mouse pad.

I planned to use a standard, stainless steel hose clamp to mount the Vantec to the base plate. I cut the clamp in half with some tin snips and straightened out the two pieces. Next, I drilled two, 1/4-inch bolt holes on the ends of the two pieces—two holes keep the clamp end from spinning around when tightening the bolts. To mount the clamp pieces to the base plate, I put the clamp pieces in a vise and used a hammer to make them form 90-degree angles that matched the sides of the Vantec.

If you use a standard hose clamp in this way, you may have problems drilling the thin metal, because once the drill is almost through, the metal will catch on the drill and twist around it. Get some scrap metal that is about an 1/8-inch thick and drill the hole in it first. Next, put the piece of clamp on a scrap piece of sturdy wood. Then, center the hole in the scrap metal over where you want the hole in the clamp end. Use Vise-Grips or C-clamps to hold the scrap metal in place, pinching the end of the hose clamp between the wood and

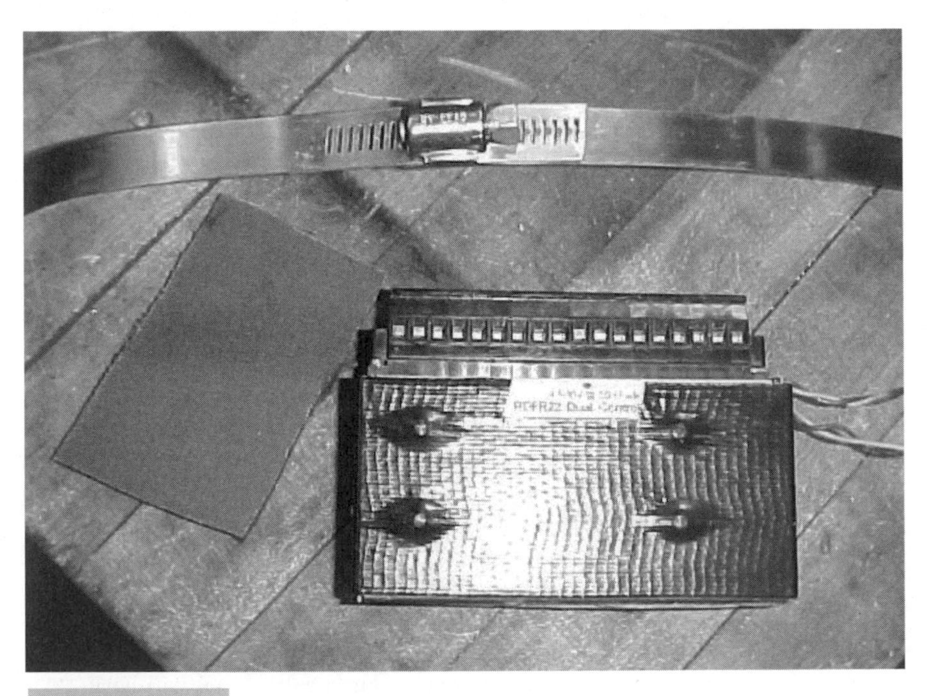

FIGURE 21.30 **Vantec, hose clamp, and mouse pad.**

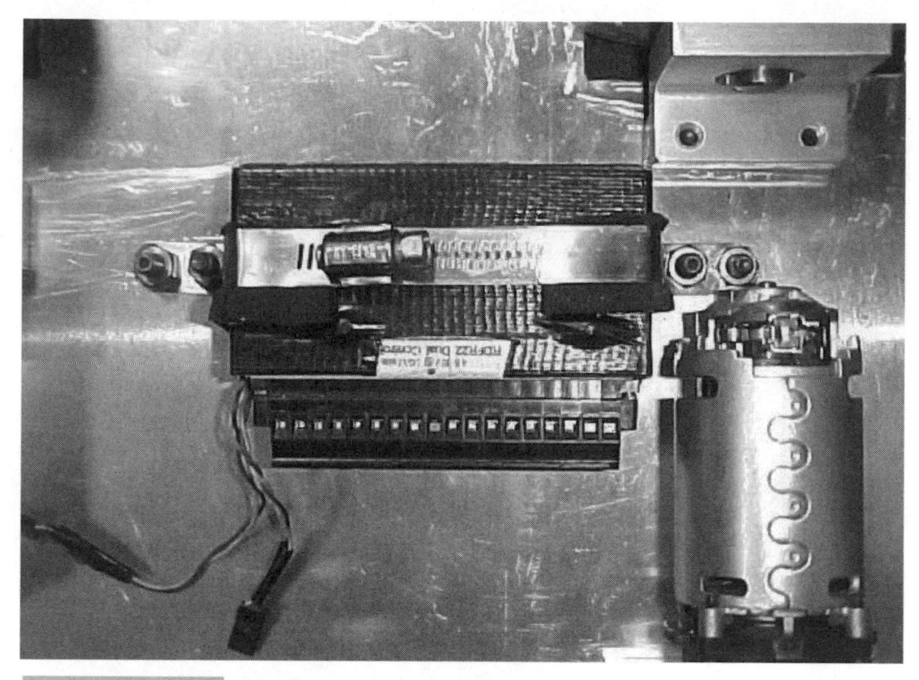

FIGURE 21.31 **Vantec mounting with a hose clamp.**

the scrap. This will hold it flat and keep it from twisting on the drill when you put the holes in it.

The large piece of mouse pad goes between the base plate and the Vantec. I used hot glue to hold the mouse pad to the Vantec, but that is not necessary. The two small pieces of mouse pad are meant to protect the sides of the controller from the metal clamp, as shown in Figure 21.31.

Connections. The RDFR23 has several different connection points that you must hook up for your bot to work. You must do it properly, too. If you connect your motors incorrectly, you will not be able to steer your bot. If you connect the power incorrectly, the unit is sure to burn up. Every RDFR23 has four connectors for the supply ground and two for the supply positive. They also have two connectors each for the first and second motors' positive and negative leads. There is one "no connect" connector. There are two connectors for connection to a braking relay.

In Dagoth, we are not using the braking function. Tighten the braking and the "no connect" screws down so that they do not rattle around. As I've said, each major connection has two connector points, so that the connectors can spread out the amount of current they have to handle. It is sort of like parallel wired batteries. To be safe, you must connect both connector points to the wire. There is an easy way to accomplish this using 10 AWG fork terminal ends bent slightly to fit into both connector points at one time. Smaller gauge terminal ends and wire can be used, but I went with 10 AWG. Each wire to each motor will have a slightly bent fork terminal end. Each wire to the battery will have a slightly bent fork terminal end. I used three fork terminals to connect all four supply ground connector

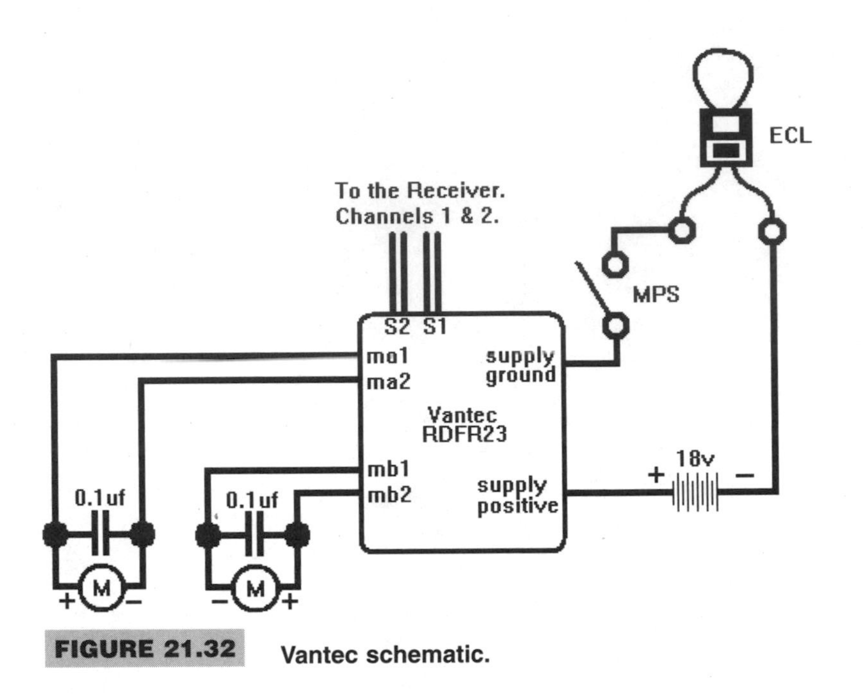

FIGURE 21.32 Vantec schematic.

points to each other. First, I clipped off the crimp side of the terminal so that they would not be in the way. Using a small screwdriver, I tightened all the connector screws. Figure 21.32 shows the electrical connections of the Vantec speed controller.

The master power switch (MPS) and the emergency cutoff loop (ECL) are shown as well. Chapter 1 explained their uses and how to connect them. In Dagoth, I used a 75-amp MPS that I had in the junk box. This particular MPS is the "pull on–push off" type. I needed access to that switch from the outside of the robot. Using two small blocks of scrap polycarbonate, I constructed a switch mount that could be bolted to the front piece of aluminum armor. The armor needed three holes drilled in it. Two holes accommodate the mounting bolts and the third, or center, hole is for access to the switch.

To use this switch in this manner, I had to make a special key. The key is made from 1/4-inch round steel. On a lathe, I drilled a hole in the end of the key and then cut threads in the hole using a tap that matched the threads of the switch handle. To activate the bot, you must put the key in the hole and turn a few revolutions. Pull on the key to turn the switch on and then unscrew the key. To deactivate the bot, you simply use the other end of the key to push the switch back in. With whatever method or switch you use, be sure you can activate and deactivate your bot within about 30 seconds. Some competitions require it.

The ECL was mounted in much the same way as the switch that supplies power to the receiver. I used a small block of scrap polycarbonate as a mount. In another robot, I used the Anderson PowerPole connectors as a small ECL. I used the same connector for Dagoth, but I only needed half of what is shown in Figure 21.33. Because the PowerPole connectors form a small hole when mounted together, I could use a roll pin to secure the connector to the polycarbonate mounting block. The pin restrained the connector so that it could not move side to side or up and down. However, the connector could easily slide forward off the pin; I used a wire tie and some hot glue to keep this from happening.

FIGURE 21.33 Dagoths' emergency cutoff loop.

Two more connectors on the speed controller bear thinking about. The two receiver input lines on my Vantec have two colored wires each. One is a ground connection and is brown; the other is the signal connection and is yellow. My Vantec came second-hand and did not have markings to designate which channel was number one and which channel was number two. I eventually had to figure it out through trial and error.

To determine, by trial and error, which signal lead should be on which receiver channel, you must connect batteries and motors to the speed controller. Because you know that your drive motors are working, you can test using these. Be sure to raise the wheels of the bot off the ground or work bench. Make sure the bot will not tilt and make the wheels touch the ground or bench. Connect everything as shown in Figure 21.34.

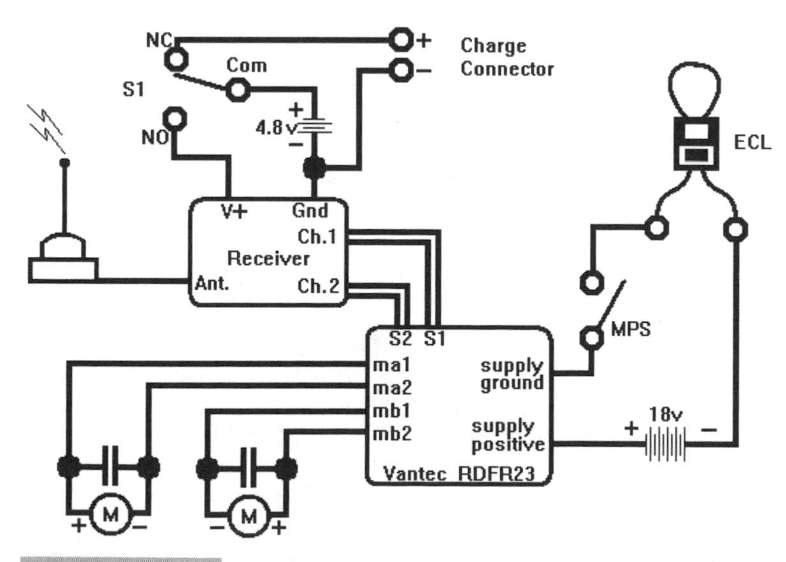

FIGURE 21.34 Complete electrical system for Dagoth.

Follow the startup rules outlined in Chapter 1 and restated here.

1. Turn on the transmitter.
2. Make sure all the trim settings are correct and joysticks or wheels are centered.
3. Make sure all the switches are set so that active weapons are deactivated.
4. Turn on the receiver and wait several seconds for it to settle.
5. Turn on the main power to the robot.

Now that the bot is powered up, move the transmitter joystick slightly to see if any motors move. If nothing happens at all, skip to the Troubleshooting section of this chapter. Vantec controllers come from the factory with the mixing option enabled. Moving the transmitter stick forward should make both motors spin forward. If they both spin in reverse, you can either switch the connection of the motor leads at the controller or switch the connections of the motor leads at the motors. Doing this reverses the polarity and makes the motor spin in the other direction. If one motor spins forward and the other spins backward, simply change the polarity of the connections on the motor spinning backward.

Now that the motor connections are correct, try moving the stick to the right. The left-hand motor should spin forward and the right-hand motor should spin backward. This would cause the bot to spin to the right if it were on the floor. The opposite should be true if you move the stick to the left. If the motors do not spin the way they should, switch the receiver connections and start over. If the motors do not spin in a manner that would cause the bot to turn, or do not spin at all when the stick is moved to the right or left, recheck the motor wiring to the controller. It is possible to connect the motors incorrectly, so that they will not turn the bot but they will make it go forward and reverse.

TROUBLESHOOTING

If the motors do not move at all when you power up and test the bot, there is a problem. Some simple troubleshooting is in order.

Make sure both battery packs are fully charged. Recheck your wiring in all systems. Recheck the way you have your receiver connected to the speed controller and its battery pack. Make sure you are getting power to the receiver. Use a servo to check that your receiver and transmitter are working and that each channel is working. Make sure you are getting good connections at the receiver end—a bent pin may not complete the required connection.

If you still cannot find the problem, go back to Chapter 1 and look at some of the simple examples again. Use the simple example circuits to test that your motors will turn with just a battery. If not, try using a new battery pack. If the new battery pack works, try out the entire system again. If you still have problems, go back to the simple circuit of the motor and battery. Put your MPS in the line and see that it is operating correctly. Next, put your ECL in the line. If everything works until you connect your speed controller, recheck that wiring again. Nine times out of ten you will find the problem in the wiring. The other one time it will be a problem inside your controller. If you have a spare controller, use it. If you do not have a spare, contact the manufacturer about repair or replacement.

One time, I triple checked the wiring to a certain interface and still did not find the problem. I emailed the manufacturer and he had me check again. On the fourth try, I found the problem. The moral is that you should put it down for a while and come back to it. If you are tired from working into the middle of the night, you will miss something that you would ordinarily catch.

A lot of bots, when turned on for the first time, tend to creep forward or move in the wrong direction. This is usually because your trim settings are not correct. The trim for one channel is the small slide either beside or below the joystick on your transmitter. Moving the lever one way or the other changes the center point of your joystick. Make sure these levers are centered and adjusted so that the bot is quiet and not moving when you let go of the transmitter stick.

WIRING

Two sections of wiring exist inside Dagoth. The wires that connect the receiver to its battery, switch, charger, and the speed controller are small and do not carry much current. The wires that connect the speed controller to the motors, MPS, ECL, and battery pack are 10 AWG, and conceivably carry between 40 and 70 amps, depending on the quality of the NiCds in the battery pack. (I could probably have used 12 AWG, but I did not have any of that size and I tend to use what I have on hand instead of running out to buy brand new supplies for every robot.)

I used the Anderson PowerPole brand connectors in several points of the schematic. I used connectors at the motors so that they could be switched out without messing with the speed controller. As I said while discussing the ECL, these connectors form a hole in the center when mounted together. I use small wire ties, threaded through these holes, to make sure the connectors will not come apart unless I want them to.

I already mentioned installing capacitors on the motors to reduce RFI problems. Following another suggestion from Chapter 2, twist the power leads to the motors and batteries to help combat RFI. (Doing this after you cut the wire can get tricky if you did not leave enough slack.)

Many competitions now require that there be no bare connections that conduct electricity. This means anything and everything that carries current. Most small connectors, as on the receiver, are excused but the large ones carrying lots of amps must be insulated. In Dagoth, three points are bare and carry lots of amps.

The MPS I used has two exposed terminals. In this instance, I simply wrapped the contacts in electrical tape. It is also possible to "paint" them with the Grip-Guard® we used to cover the contacts of the motor capacitors, and some people will go as far as using the Grip-Guard® and electrical tape.

The second bare point is where the motor's brushes connect to the motor power leads. I used a female spade connector to slide down over the male spade connector in the DeWalt brush housing. Although the connections are on opposite sides of the motor, some safety inspectors still insist that these be covered. If this is the case with your inspector, use high temperature, silicone tape to cover these connections. The motors get hot enough to make regular electrical tape lose its grip.

The third and most important bare point is at the speed controller connector housing. The fork terminals we used to connect the leads to the controller are bare metal. Some builders use electrical tape to insulate these connections, others use the Grip-Guard. I do not like putting Grip-Guard® on my $200+ piece of equipment, because is difficult to remove it when you need to.

Weapons

The weapons I chose for Dagoth are simple and inactive—three, hardened spikes that I hope will penetrate light armor and tires. The wedge will be used to get under our opponents and push them around.

The spikes were made from large, grade 5 bolts. First, I cut the head off the bolt with the bandsaw. Next, the headless bolt was chucked in a lathe and turned into a spike. The threads of the bolt were saved for mounting the spikes. The spike shape I chose was that of a center punch. One very short section on the tip of the spike is about 45 degrees, forming tip or point of the spike. I did not want a very sharp spike point because it tends to get bent easily.

The spikes were mounted into the front aluminum armor. Three holes were drilled into the armor. A thin nut was run onto the threads and up to the unthreaded portion, then the spike was inserted into the armor, and another nut was used to tighten the spike down.

The wedge was made from 16 AWG stainless steel. I used the bandsaw to cut the rectangle. Next I picked up a long stainless steel hinge so that I could mount the wedge onto the rear aluminum armor. The hinge makes the wedge mobile. Mobility is important for a couple reasons. The first reason is that a solidly mounted wedge can get stuck on seams in the arena or can get bent and cause your wheels to leave the ground. The second reason is that your bot can get flipped upside down. If the wedge could not move, it would be useless. Because of the hinge, the wedge can flip down to the floor whenever the bot gets inverted. Mounting the hinge on the aluminum armor is simple. Mark the existing holes using a hole transfer. Drill and tap the marked holes. Mark the existing holes on the wedge material using the hole transfer. Drill the holes in the wedge material. Bolt it all together, and the bot has its second weapon. Figure 21.36 shows the completed Dagoth tipping the scales at 29.5 pounds. RAID is sitting on top of him.

FIGURE 21.35 Installed spikes.

FIGURE 21.36 **Finished Dagoth with the wedge and spikes.**

Cost Rundown

That's about it for Dagoth. The only thing left are the costs. Let me run them down for you real quick:

Aluminum base plate, armor, scrap	$25.00
24" × 24" × 0.25" polycarbonate	$15.00
12" × 24" × 0.5"" polycarbonate	$31.00
Spike and wedge material	$20.00
Hitech Prism remote control system	$412.00
Base loaded antenna	$11.00
DeWalt 18-volt battery	$80.00
DeWalt 18-volt battery charger	$60.00
Master Power Switch	$4.00
25 Red and 25 black Anderson Powerpole connectors	$20.00
6 Feet of red and 6 feet of black 10 AWG stranded wire	Scrap

continued on next page

25 10 AWG fork connectors	$3.00
25 10 AWG spade connectors	$3.00
2 0.1-μf capacitors	$1.00
Vantec RDFR23	$340.00
Large radiator hose clamp	$2.00
Old mouse pad	$ Free
Teflon scrap	$ Free
Electrical tape	$2.00
Grip Guard®	$10.00
Two 18-V DeWalt motors, transmissions, Team Delta mounts, axles	$290.00
12 inches of 0.5-inch keyed axle	$10.00
Four Team Delta small pillow block bearings	$98.00
Four 4-inch Colson wheels with hubs from Team Delta	$208.00
Four idler sprockets	$56.00
10 feet of #35 roller chain	$20.00
6 inches of 1/8-inch key stock	$0.50
2 half links	$2.40
2 master links	$1.00
100 pc. box of quarter-twenty by 1-inch bolts	$16.00
100 pc. box of quarter-twenty by 3/4-inch bolts	$16.00
25 pc. box of 10-24 by 1-inch bolts	$4.00
100 wire ties	$5.00
Total	**$1765.90**

Summary

This chapter offers a complex yet brief view of how I built the robot named Dagoth. When I set out to write it, I had many ideas that I wanted to cover. The problem was that if I covered every single step of building this robot, the book would have been too big and no one would have read the whole thing. Instead of a paint-by-numbers robot, I've outlined the major steps and tossed in several lines of thought that I was following. Hopefully, I included enough to help you construct your own metal monster.

The next chapter will aid you in getting your robot running if you have problems. We also discuss practice methods and how to strengthen your robot's frame and armor. Just in case your bot is overweight, we'll talk about putting it on a crash diet.

22

TWEAKING

There are two keys to winning robot combat matches. One is building a strong bot. The other is driving practice. There are a few different levels of driving practice. First, you want to get the feel of how the bot moves. Then, you need to practice with some goals in mind. Dogfights are last, but they are also the most fun. Building a strong bot takes a mixture of knowledge and testing. Hopefully, this book contains enough knowledge to get you to the testing part.

Practice

You should spend lots of time driving your robot. Of course, you can only practice driving your robot if you finish building it before the competition. I once took a robot I had never driven before to a competition in Las Vegas. The reason I had never driven it was that I only finished building it the night before we got on the airplane. That night, we had to take the robot apart so that we could bring it out with us in our luggage. After that competition, my team and I promised ourselves that we would never again go to a competition without a finished bot. In fact, we promised that we would get the bot done a month ahead of time. With the exception of Dagoth, we have not been able to get a bot done a full month ahead of time, because there are always so many things going on that two weeks is the most we could ever muster. Of all the competitions that we attended after Vegas, we brought an unfinished bot to only one. In our defense, we were planning the second NC Robot

StreetFight and building its arena. The bot was overweight without any weapon or armor, so we were bumped into the next higher weight class. When it was our turn to fight, we had radio problems and ground clearance problems. Nothing is more embarrassing than putting a robot that does not work in the ring. If you do not have a finished bot when you get to the competition, you automatically lose the second key to winning. You must spend time driving your robot to win.

The first kind of practice you should get is meant to get you used to how the bot moves. You have to get used to its speed, turning ability, aiming ability, and stopping ability. Smaller bots can be driven around in a carport or on a cement driveway. Larger bots need more room. If you live in a development where there is not much traffic, you may be able to practice on the street. I live on a busy road, so I take my large bots to an empty parking lot. Practice driving away and returning. Practice driving side to side. Spend time doing lots of turns and spins. Do some 90- and 180-degree turns. Put the driving and the turns together. Spend lots of time running from side to side in the same line and turning on the same spots. Spend time with 90-degree turns by driving in a square or rectangle around yourself. Challenge a brick wall. Don't slam into it but charge straight at it and hit the brakes. See how close you can come to it without making contact.

Once you get the hang of driving your bot, it is time to add some obstacles. In a real competition, there will likely be some type of obstacle in the arena that you want to avoid. Whether saws in the floor, hammers in the corners, or flame pits, you want to avoid whatever is in the ring besides your opponent. Use traffic cones to set up a driving course. Practice driving the course while standing in different spots. Once you feel comfortable with driving the course, start trying to push an old tire through the course. If you have a smaller bot, use a smaller tire or a cardboard box. Once you can push the tire or box around, it is time to graduate to something more difficult. Try pushing a basketball (or ball of appropriate size compared to the robot) through the course. The ball will get away from you really easily. Do not worry about keeping the ball in front of the bot at first. Simply get the ball through the course. Continue practicing while standing in different spots around the course. If you get good at pushing a round ball through, start working with an American football.

All the practicing you have done so far is preparing you for a robot fight. It is getting you used to driving the bot and to anticipating which direction you will need to be moving. The next type of driving is much like the dogfights the Armed Forces pilots practice. In this case, you will be fighting a toy remote control car. Get a friend to drive the RC car. Practice chasing the car without any weapons activated unless you want to buy new toys every time you practice. Practice running from the car. Sometimes you will be up against an opponent where the best strategy is to avoid it until you get a good shot at using your own weapon.

Strength

Once you get good at dogfights with the toys, it is time to start testing the toughness of your metal warrior. Most people who build robots with steel or aluminum tube frames will test the frame before they mount any motors or electronics. If you have a frame of that type, you can test its durability by standing on it. You can also drop it from different heights onto concrete or asphalt surfaces. I would not recommend running over it with a car, but that is up to you. If your frame is damaged or warped from the tests, it is not strong enough. You will be amazed at the power of your opponents when inside the arena. There is not much you can do to your frame that will cause the same amount of damage as your opponent, so be ruthless when testing it. Repair what needs to be repaired. Add gussets to corners that did not hold their shape. Make sure there are no cracked welds. If any bolts break, pay attention to how they were used. If they were in a position to shear, determine how to avoid using the bolts in the same manner while accomplishing the same task. If they weren't in a position to shear, use bigger bolts.

Remember the brick wall driving test? Now is the time to return to it. Instead of stopping, slam into it. Do it several times at full speed. If nothing breaks, your machine is ready for combat. This may seem like an extreme way to test, but up until now we have had no way of knowing whether or not the drivetrain was actually strong enough to withstand battle. We also have had no way to determine if our electronics were mounted so that any impact shock would not affect them. Replace whatever needs replacing. Repair whatever needs repairing. Change the way you mount electronics and drive components, if necessary. Finally, remember that no matter how harshly you treat your robot while testing, it's nothing compared to what it will face when in the ring with a real opponent.

Weight

You should be weighing your bot all through the building process. However, scales are sometimes difficult to gain access to or even find. My team ended up buying a set of digital scales capable of weighing up to 400 pounds for use at the NC Robot StreetFight. So, we get to weigh our bots any time we feel like it. If you don't want to buy scales, there are some places you might find them. Every manufacturing company I have ever been to has a large scale that is used for weighing whatever they ship out to the customer. The post office or other mailing companies have smaller scales that can be used to weigh smaller bots. Schools have scales for athletes. Machine shops usually have scales for weighing materials. Scrap yards have scales for weighing incoming and outgoing scrap.

Once you know what your robot weighs, you might have to put it on a crash diet. As I mentioned earlier, there are lots of places on a bot that can be lightened. If you are only a few ounces over the limit, you may be able to lose that by removing washers and the ends of bolts that are too long. If you are only a couple pounds over the limit, you have several avenues to try.

If you are using axles or weapon shafts that are 1 inch or more in diameter, you can put them in a lathe and bore them out to lose weight. Just like in the Six Million Dollar Mouse, you can shave some weight off by lightening any gears you may be using. You can put holes in lots of thick, heavy metal without losing integrity. If the axles or shafts aren't spinning too fast, you can replace heavy ball bearings with Oil-Lite style bearings. If you can't replace the bearing itself, you may be able to find a lighter version. Lots of pillow block and flange bearings come in aluminum as well as steel or cast iron. Aluminum is usually a little more expensive, but it will help keep the weight down.

If you have a spinning weapon, you know that a heavier spinner equals a harder hit on the opponent. That's bad news if you have to drop some weight. The secret is to leave as much weight as far away from the center of the spin as you can. If you're spinning a bar, you can drill or cut holes starting at the center and going outward. This can cause a strength problem and you might end up with a bent spinning bar. If you are spinning the entire shell of a robot, you may be able to remove most of the inner part and replace it with a lighter material. You may be able to leave it out all together. Either of these solutions can cause the same weakening around the spinning axis. Leaving the material out all together can leave the insides of your bot exposed to danger. Cutting holes in a spinning bar, spinning disk, or a spinning shell can throw it out of balance. Be sure you cut equal amounts of weight all over or your bot might vibrate across the arena.

Batteries can be heavy, too. While doing test runs, you should be running a timer to see how much runtime you get. If you're getting more than 5 minutes, you can lose some weight by changing batteries. If you're using SLAs, you can try smaller ones or you can try the lighter NiCds or NiMhs.

The last resort of losing weight is drilling holes in the frame, gussets, and armor. Depending on the type of materials you use, you can expect to drill lots and lots of holes to lose a few ounces. Ounces may be what you need after going through all the other steps. However, if you drill too many holes, you may just weaken the part that helps make your weapon effective or your bot durable. Be careful about what you take out.

Summary

If your plan is to enter the big competitions where prize money, TV time, and toy deals are at stake, the absolute best way to practice is against real opponents in real matches. Attend every small competition that you hear about. Check out other robots and learn from other builders' successes and failures. If you lose, determine what went wrong. Change it if the problem was mechanical, and practice if he out drove you. If you win, do not dwell on it. There is always someone stronger or better. Most of all, build strong, build mean, be safe, and have fun.

DO'S AND DON'TS

Do... Document everything you do. You will not remember it all.

Do... Figure out which size of wire you will need and use it.

Do... Try to find some wire with heat-resistant insulation.

Do... Keep wires neat but leave some extra length for adjustments.

Do... Cover battery terminals with some type of insulator.

Do... Tape or shrink-wrap all wire connectors.

Don't... Use a lot of connectors. They create failure points in wiring.

Do... Shock-mount everything to reduce vibration to electronics.

Do... Use capacitors on motor brushes to reduce RC interference.

Do... Twist all power leads to reduce RC interference.

Do... Mount your antenna inside some Lexan or the bot itself to protect from attacks.

Do... Be conscious of your transmitter and receiver battery's charge level.

Do... Get the right size speed controller for your motors.

Do... Keep your batteries charged.

Do... Have enough batteries to change them between matches.

Do... Mount motors solidly to the chassis for stability.

Do... Grease gears, chains, and sprockets for best performance.

Do... Use a thread locker solution to hold critical bolts and setscrews.

Do… Use keyed shafts and gears, sprockets, or pulleys instead of roll pins.

Do… Use square shafts and square-holed gears, sprockets, or pulleys instead of keyed shafts.

Do… Use layers of materials instead of one thick piece.

Do… Devise a plan and try to stick to it.

Do… Keep It Simple Stupid.

Don't… Enter the event if your bot isn't finished. Pay for a ticket and be a spectator.

Do… Practice driving your robot as much as possible.

Do… Know how your opponent works and try to take advantage of his weaknesses.

Do… Be conscious of your surroundings in the pit area.

Do… Be courteous to other builders in the pit area.

Do… Return borrowed tools as soon as you are finished with them.

Don't… Take anyone's word that something won't work. Build a model and see.

Do… Wear protective gloves, face shields, and goggles.

Don't… Wear loose clothing or jewelry in the shop or while operating the bot.

Do… BE SAFE and have fun!

B

CONVERSION FACTORS

DISTANCE

1 millimeter (mm) = 0.03937 inches (in)

1 inch (in) = 25.4 millimeters (mm)

1 centimeter (cm) = 10 millimeters (mm)

1 centimeter (cm) = 0.3937 inches (in)

1 inch (in) = 2.54 centimeters (cm)

1 meter (m) = 3.2808 feet (ft)

1 foot (ft) = 0.3048 meters (m)

1 kilometer (km) = 0.62137 miles (mi)

1 mile (mi) = 1.6093 kilometers (km)

1 mil (mil) = 0.001 inches (in)

1 inch (in) = 1000 mils (mil)

1 foot (ft) = 12 inches (in)

1 inch (in) = 0.083333 feet (ft)

1 yard (yd) = 3 feet (ft)

1 yard (yd) = 36 inches (in)

1 yard (yd) = 36000 mils (mil)

AREA

1 square centimeter (cm^2) = 0.155 square inches (in^2)

1 square inch (in^2) = 6.4516 square centimeters (cm^2)

1 square meter (m^2) = 10.764 square feet (ft^2)

1 square foot (ft^2) = 0.0929 square meters (m^2)

1 square meter (m^2) = 1.196 square yards (y^2)

1 square yard (y^2) = 0.8361 square meters (m^2)

1 square foot (ft^2) = 144 square inches (in^2)

1 square inch (in^2) = 0.00694 square feet (ft^2)

1 square yard (y^2) = 9 square feet (ft^2)

1 square foot (ft^2) = 0.1111 square yards (y^2)

1 square yard (y^2) = 1296 square inches (in^2)

VOLUME

1 cubic centimeter (cm^3) = 0.0610 cubic inches (in^3)

1 cubic inch (in^3) = 16.387 cubic centimeters (cm^3)

1 cubic meter (m^3) = 35.315 cubic feet (ft^3)

1 cubic foot (ft^3) = 0.0283 cubic meters (m^3)

1 cubic meter (m^3) = 1.308 cubic yards (y^3)

1 cubic yard (y^3) = 0.7646 cubic meters (m^3)

1 cubic foot (ft^3) = 1728 cubic inches (in^3)

1 cubic inch (in^3) = 0.0005787 cubic feet (ft^3)

1 cubic yard (y^3) = 27 cubic feet (ft^3)

1 cubic foot (ft^3) = 0.03704 cubic yards (y^3)

1 cubic yard (y^3) = 46656 cubic inches (in^3)

1 gallon = 231 cubic inches (in^3)

1 gallon = 0.13368 cubic feet (ft^3)

1 gallon = 3785.4 cubic centimeters (cm^3)

1 gallon = 3.7854 liters

1 gallon = 3785.4 milliliters

1 gallon = 4 quarts

1 gallon = 8 pints (pt)

1 gallon = 16 cups

1 gallon = 128 ounces (oz)

1 gallon = 256 tablespoons

1 gallon = 768 teaspoons

TEMPERATURE

1 degree celsius (C) = 33.8 degrees fahrenheit (F)

1 degree fahrenheit (F) = −17.222 degrees celsius (C)

0 degrees celsius (C) = 32 degrees fahrenheit (F)

0 degrees fahrenheit (F) = −17.778 celsius (C)

1 degree celsius (C) = 274.15 degrees kelvin (K)

1 degree kelvin (K) = −272.15 degrees celsius (C)

1 degree kelvin (K) = −457.87 degrees fahrenheit (F)

0 degrees kelvin (K) = −273.15 degrees celsius (C) (absolute zero)

0 degrees kelvin (K) = −459.67 degrees fahrenheit (F)

PRESSURE

1 atmosphere = 14.696 pounds per square inch (psi)

1 atmosphere = 2116.2 pounds per square foot

1 atmosphere = 1.0332 kilograms per square centimeter

1 atmosphere = 10332 kilograms per square meter

1 atmosphere = 1.0133 bar

ENERGY

1 Joule (J) = 0.00094782 Brittish Thermal Units (BTU)

1 Joule (J) = 0.23885 calories

1 Joule (J) = 3.7251e-007 horsepower-hour

1 Joule (J) = 1 newton-meter (nm)

1 Joule (J) = 141.61 ounce force-inch (ozf)

1 Joule (J) = 0.00027778 watt-hours

1 Joule (J) = 1 watt-second

POWER

1 horsepower (hp) = 2546.7 BTU per hour

1 horsepower (hp) = 0.7457 kilowatt (kw)

1 horsepower (hp) = 550 pound-feet per second

1 horsepower (hp) = 745.7 watts

1 kilowatt (kw) = 1000 watts

1 kilowatt (kw) = 737.56 pound-feet per second

1 kilowatt (kw) = 1.341 horsepower (hp)

MASS

1 ounce (oz) = 28.35 grams (g)

1 ounce (oz) = 0.02835 kilograms (kg)

1 ounce (oz) = 28350 milligrams (mg)

1 pound (lb) = 453.59 grams (g)

1 pound (lb) = 0.45359 kilograms (kg)

1 pound (lb) = 4.5359e+005 milligrams (mg)

1 pound (lb) = 16 ounces (oz)

1 pound (lb) = 0.0005 tons (tn)

1 ton (tn) = 2000 pounds (lb)

1 ton (tn) = 0.90718 tonnes (metric ton)(t)

1 tonne (metric ton) (t) = 1.1023 tons (tn)

1 tonne (metric ton) (t) = 2204.6 pounds (lb)

FORCE

1 pound-force (lbf) = 16 ounce-force (ozf)

1 pound-force (lbf) = 4.4482 newtons (N)

1 pound-force (lbf) = 0.45359 kilogram-force

1 pound-force (lbf) = 453.59 grams-force

1 newton (N) = 0.22481 pound-force (lbf)

1 newton (N) = 3.5969 ounce-force (ozf)

1 newton (N) = 0.10197 kilogram-force

1 newton (N) = 101.97 grams-force

SPEED

1 mile per hour (mph) = 1.6093 kilometer per hour (kph)

1 mile per hour (mph) = 5280 feet per hour

1 mile per hour (mph) = 88 feet per minute

1 mile per hour (mph) = 1.4667 feet per second

1 mile per hour (mph) = 0.000947 inches per minute

1 kilometer per hour (kph) = 0.62137 miles per hour (mph)

1 kilometer per hour (kph) = 3280.8 feet per hour

1 kilometer per hour (kph) = 54.681 feet per minute

1 kilometer per hour (kph) = 0.91134 feet per second

1 knot = 1.1508 miles per hour (mph)

1 knot = 1.852 kilometers per hour (kph)

1 knot = 0.0015521 mach

1 mach = 741.45 miles per hour (mph)

1 mach = 1193.3 kilometers per hour (kph)

ANGLE

1 degree = 1.1111 grads (g)

1 degree = 0.017453 radians (rad)

1 degree = 60 minutes (')

1 degree = 3600 seconds ('')

1 degree = 0.0027778 revolutions (r)

1 grad (g) = 0.9 degrees

1 grad (g) = 0.015708 radians (rad)

1 grad (g) = 54 minutes (')

1 grad (g) = 3240 seconds ('')

1 grad (g) = 0.0025 revolutions (r)

1 radian (rad) = 57.296 degrees

1 radian (rad) = 63.662 grads (g)

1 radian (rad) = 3437.8 minutes (')

1 radian (rad) = 2.0627e+005 seconds ('')

1 radian (rad) = 0.15915 revolutions (r)

1 minute (') = 0.016667 degrees

1 minute (') = 0.018519 grads (g)

1 minute (') = 0.00029089 radians (rad)

1 minute (') = 60 seconds ('')

1 minute (') = 4.6296e-005 revolutions (r)

1 second ('') = 0.00027778 degrees

1 second ('') = 0.00030864 grads (g)

1 second ('') = 4.8481e-006 radians (rad)

1 second ('') = 0.016667 minutes (')

1 second ('') = 7.716e-007 revolutions (r)

INFORMATION TABLES

TABLE C.1 AMERICAN RC CHANNEL FREQUENCIES

27 MHZ BAND (AIRCRAFT/CAR/BOAT)
26.995 MHz—Chan. 1—Brown
27.045 MHz—Chan. 2—Red
27.095 MHz—Chan. 3—Orange
27.145 MHz—Chan. 4—Yellow
27.195 MHz—Chan. 5—Green
27.255 MHz—Chan. 6—Blue
50 MHZ BAND (AIRCRAFT/CAR/BOAT) FCC AMATEUR LICENSE REQUIRED.
50.800 MHz—Chan. RC00
50.820 MHz—Chan. RC01
50.840 MHz—Chan. RC02
50.860 MHz—Chan. RC03
50.880 MHz—Chan. RC04
50.900 MHz—Chan. RC05
50.920 MHz—Chan. RC06
50.940 MHz—Chan. RC07
50.960 MHz—Chan. RC08
50.980 MHz—Chan. RC09
72 MHZ BAND (AIRCRAFT ONLY)
72.010 MHz—Chan. 11
72.030 MHz—Chan. 12
72.050 MHz—Chan. 13
72.070 MHz—Chan. 14
72.090 MHz—Chan. 15
72.110 MHz—Chan. 16
72.130 MHz—Chan. 17
72.150 MHz—Chan. 18
72.170 MHz—Chan. 19
72.190 MHz—Chan. 20
72.210 MHz—Chan. 21

continued on next page

TABLE C.1 AMERICAN RC CHANNEL FREQUENCIES (continued)

72 MHZ BAND (AIRCRAFT ONLY) (CONTINUED)

72.230 MHz—Chan. 22
72.250 MHz—Chan. 23
72.270 MHz—Chan. 24
72.290 MHz—Chan. 25
72.310 MHz—Chan. 26
72.330 MHz—Chan. 27
72.350 MHz—Chan. 28
72.370 MHz—Chan. 29
72.390 MHz—Chan. 30
72.410 MHz—Chan. 31
72.430 MHz—Chan. 32
72.450 MHz—Chan. 33
72.470 MHz—Chan. 34
72.490 MHz—Chan. 35
72.510 MHz—Chan. 36
72.530 MHz—Chan. 37
72.550 MHz—Chan. 38
72.570 MHz—Chan. 39
72.590 MHz—Chan. 40
72.610 MHz—Chan. 41
72.630 MHz—Chan. 42
72.650 MHz—Chan. 43
72.670 MHz—Chan. 44
72.690 MHz—Chan. 45
72.710 MHz—Chan. 46
72.730 MHz—Chan. 47
72.750 MHz—Chan. 48
72.770 MHz—Chan. 49
72.790 MHz—Chan. 50
72.810 MHz—Chan. 51
72.830 MHz—Chan. 52
72.850 MHz—Chan. 53
72.870 MHz—Chan. 54
72.890 MHz—Chan. 55
72.910 MHz—Chan. 56
72.930 MHz—Chan. 57
72.950 MHz—Chan. 58
72.970 MHz—Chan. 59
72.990 MHz—Chan. 60

75 MHZ BAND (CAR/BOAT ONLY)

75.410 MHz—Chan. 61
75.430 MHz—Chan. 62
75.450 MHz—Chan. 63
75.470 MHz—Chan. 64
75.490 MHz—Chan. 65
75.510 MHz—Chan. 66
75.530 MHz—Chan. 67
75.550 MHz—Chan. 68
75.570 MHz—Chan. 69
75.590 MHz—Chan. 70
75.610 MHz—Chan. 71
75.630 MHz—Chan. 72
75.650 MHz—Chan. 73
75.670 MHz—Chan. 74
75.690 MHz—Chan. 75
75.710 MHz—Chan. 76
75.730 MHz—Chan. 77
75.750 MHz—Chan. 78
75.770 MHz—Chan. 79
75.790 MHz—Chan. 80
75.810 MHz—Chan. 81
75.830 MHz—Chan. 82
75.850 MHz—Chan. 83
75.870 MHz—Chan. 84
75.890 MHz—Chan. 85
75.910 MHz—Chan. 86
75.930 MHz—Chan. 87
75.950 MHz—Chan. 88
75.970 MHz—Chan. 89
75.990 MHz—Chan. 90

TABLE C.2 CHART OF SERVO LEAD COLORS BY MANUFACTURER*

SERVO	POSITIVE (+)	SIGNAL (S)	NEGATIVE (−)
Futaba - J	Red	White	Black
JR	Red	Orange	Brown
Hitec	Red	Yellow	Black
Airtronics	Red	Orange	Black
	Red	White	Black
	Red	Black	Black
Airtronics - Z	Red	Blue	Black
Fleet	Red	White	Black
KO	Red	White	Black

*Robert's Gadgets & Gizmos.

TABLE C.3 TABLE OF DRILL SIZES AND THREADS (DERIVED FROM TC 9-524)

DRILL SIZE	DECIMAL	THREAD	DRILL SIZE	DECIMAL	THREAD
80	.0135		7	.2010	1/4-20
79	.0145		13/64	.2031	
1/64	.0156		6	.2040	
78	.0160		5	.2055	
77	.0180		4	.2090	
76	.0200		3	.2130	1/4-28
75	.0210		7/32	.2188	
74	.0225		2	.2210	
73	.0240		1	.2280	
72	.0250		A	.2340	
71	.0260		15/64	.2344	
70	.0280		B	.2380	
69	.0292		C	.2420	
68	.0310		D	.2460	
1/32	.0312		1/4	.2500	
67	.0320		E	.2500	
66	.0330		F	.2570	5/16-18
65	.0350		G	.2610	
64	.0360		17/64	.2656	

continued on next page

TABLE C.3 TABLE OF DRILL SIZES AND THREADS (DERIVED FROM TC 9-524) (continued)

DRILL SIZE	DECIMAL	THREAD	DRILL SIZE	DECIMAL	THREAD
63	.0370		H	.2660	
62	.0380		I	.2720	5/16-24
61	.0390		J	.2770	
60	.0400		K	.2810	
59	.0410		9/32	.2812	
58	.0420		L	.2900	
57	.0430		M	.2950	
56	.0465		19/64	.2969	
3/64	.0469	0-80	N	.3020	
55	.0520		5/16	.3125	3/8-16
54	.0550		O	.3160	
53	.0595	1-64 and 1-72	P	.3230	
1/16	.0625		21/64	.3281	
52	.0635		Q	.3320	3/8-24
51	.0670		R	.3390	
50	.0700	2-56 and 2-64	11/32	.3438	
49	.0730		S	.3480	
48	.0760		T	.3580	
5/64	.0781		23/64	.3594	
47	.0785	3-48	U	.3680	7/16-14
46	.0810		3/8	.3750	
45	.0820	3-56	V	.3770	
44	.0860		W	.3860	
43	.0890	4-40	25/64	.3906	7/16-20
42	.0935	4-48	X	.3970	
3/32	.0938		Y	.4040	
41	.0960		13/32	.4062	
40	.0980		Z	.4130	
39	.0995		27/64	.4219	1/2-12 and 1/2-13
38	.1015	5-40	7/16	.4375	
37	.1040	5-44	29/64	.4531	1/2-20
36	.1065	6-32	15/32	.4688	
7/64	.1094		31/64	.4844	9/16-12
35	.1100		1/2	.5000	
34	.1110		33/64	.5156	9/16-18

continued on next page

TABLE C.3 TABLE OF DRILL SIZES AND THREADS (DERIVED FROM TC 9-524) (continued)

DRILL SIZE	DECIMAL	THREAD	DRILL SIZE	DECIMAL	THREAD
33	.1130	6-4017/32	.5312	5/8-11	
32	.1160		35/64	.5469	
31	.1200		9/16	.5625	
1/8	.1250		37/64	.5781	5/8-18
30	.1285		19/32	.5938	
29	.1360	8-32 and 8-36	39/64	.6094	
28	.1405		5/8	.6250	
9/64	.1406		41/64	.6406	
27	.1440		21/32	.6562	3/4-10
26	.1470		43/64	.6719	
25	.1495	10-24	11/16	.6875	3/4-16
24	.1520		45/64	.7031	
23	.1540		23/32	.7188	
5/32	.1562		47/64	.7344	
22	.1570		3/4	.7500	
21	.1590	10-32	49/64	.7656	7/8-9
20	.1610		25/32	.7812	
19	.1660		51/64	.7969	
18	.1695		13/16	.8125	7/8-14
11/64	.1719		53/64	.8281	
17	.1730		27/32	.8438	
16	.1770	12-24	55/64	.8594	
15	.1800		7/8	.8750	1-8
14	.1820	12-28	57/64	.8906	
13	.1850		29/32	.9062	
3/16	.1875		59/64	.9219	1-12 and 1-14
12	.1890		15/16	.9375	
11	.1910		61/64	.9531	
10	.1935		31/32	.9688	
9	.1960		63/64	.9844	1-1/8–7
8	.1990		1	1.0000	

Decimal equivalents of fraction, wire gauge, and letter size drills.

Tap drill sizes based on approximately 75 percent Full Thread.

TABLE C.4 TAPPING LUBRICANTS CHART (DERIVED FROM TC 9-524)

MATERIAL BEING CUT	LUBRICANT
Types of Steel Cast Forged Machined Stainless Tool Chrome-moly Malleable irons	Sulfur Base Oil
Cast Iron Synthetic Resins Hard Filler	Dry—No Lubricant
Brass Bronze Copper Manganese bronze	Light Oil
Plastics (polycarbonate)	Water
Aluminum Duralumin Die Castings	Kerosene WD-40

TABLE C.5 TABLE OF SPINDLE SPEEDS FOR DRILLING (DERIVED FROM TC 9-524)

	MATERIAL AND CUTTING SPEED (FT PER MINUTE)				
DRILL DIAMETER (INCHES)	ALUMINUM	MILD STEEL 0.2–0.3 CARBON	MED STEEL 0.4–0.5 CARBON	TOOL STEEL 1.2 CARBON	STAINLESS STEEL
	300	110	80	60	50
	REVOLUTIONS PER MINUTE				
1/16	18338	6724	4883	3668	3065
1/8	9156	3362	2444	1634	1528
3/16	6108	2242	1830	1222	1018
1/4	4584	1681	1222	917	764
5/16	3666	1344	978	733	611
3/8	3064	1121	818	611	609
7/16	2622	921	699	524	437
1/2	2292	840	611	458	382
9/16	2037	747	543	407	340
5/8	1838	673	489	367	308
11/16	1665	611	444	333	273
3/4	1524	569	408	308	254
13/16	1422	521	379	288	237
7/8	1314	462	349	262	219
15/16	1221	448	328	244	204
1	1140	420	308	228	191
1 1/16	1077	395	287	218	180
1/8	1020	374	272	204	170
1 3/16	966	354	258	193	161
1 1/4	918	337	245	183	153
1 5/16	873	320	233	175	145
1 3/8	834	308	222	187	138
1 7/16	795	282	212	159	133
1 1/2	762	279	204	153	127
1 9/16	732	266	195	146	122
1 5/8	702	257	188	141	117
1 11/16	678	249	181	136	113
1 3/4	664	240	175	131	108
1 13/16	630	231	168	126	105
1 7/8	612	224	163	122	102
1 15/16	601	216	158	118	88
2	573	210	153	116	86

Rotational speed value for carbide drills is 200 to 300 percent higher than H.S.S.

TABLE C.6 CHART OF WIRE SIZE, CURRENT, AND LENGTH (DERIVED FROM RBE ELECTRONICS)

WIRE GAUGES ELECTION TABLE

CIRCUIT AMPS		WIRE GAUGE FOR LENGTH IN FEET						
6 VOLTS	12 VOLTS	3'	5'	7'	10'	15'	20'	25'
0 to 2.5	0 to 5	18	18	18	18	18	18	18
3.0	6	18	18	18	18	18	18	16
3.5	7	18	18	18	18	18	18	16
4.0	8	18	18	18	18	18	16	16
5.0	10	18	18	18	18	16	16	16
5.5	11	18	18	18	18	16	16	14
6.0	12	18	18	18	18	16	16	14
7.5	15	18	18	18	18	14	14	**12**
9.0	18	18	18	16	16	14	14	**12**
10	20	18	18	16	16	14	**12**	**10**
11	22	18	18	16	16	**12**	**12**	**10**
12	24	18	18	16	16	**12**	**12**	**10**
15	30	18	16	16	14	**10**	**10**	**10**
20	40	18	16	14	**12**	**10**	**10**	8
25	50	16	14	**12**	**12**	**10**	**10**	8
50	100	**12**	**12**	**10**	**10**	6	6	**4**
75	150	**10**	10	8	8	**4**	**4**	2
100	200	**10**	8	8	6	**4**	**4**	2

Find the current you have at the voltage on the left. Find the length of wire on the top. Move down and across until the column and row meets. The number in the joining square is the size of wire you need. The bold and underlined wire sizes are what I usually use in all but the lightest robots.

TABLE C.7 CHART OF WIRE SIZE AND CURRENT CAPACITY

AWG WIRE SIZE	APPROXIMATE AMPERAGE
18	28
16	25
14	35
12	45
10	60
8	80
6	120
4	160
2	210
0	285
00	330
000	385
0000	445

(Under 50 volts with 105 C insulation)

TABLE C.8 AWG AMERICAN WIRE GAUGE/DIAMETER/RESISTANCE (DERIVED FROM READE)

WIRE SIZE AWG	DIAMETER (MM)	DIAMETER (INCH)	RESISTANCE (OHM/KM)	RESISTANCE (OHM/1000 FT)
18	1.02	0.040	21.9	6.60
16	1.29	0.051	13.0	4.20
14	1.63	0.064	8.54	2.60
12	2.05	0.081	5.40	1.70
10	2.59	0.102	3.40	1.00
8	3.73	0.147	2.20	0.67
6	4.67	0.184	1.50	0.47
4	5.90	0.232	0.80	0.24
2	7.42	0.292	0.50	0.15
0	9.35	0.368	0.31	0.096
00	10.52	0.414	0.25	0.077
000	11.76	0.464	0.20	0.062
0000	13.26	0.522	0.16	0.049

continued on next page

TABLE C.8 AWG AMERICAN WIRE GAUGE/DIAMETER/RESISTANCE (DERIVED FROM READE) (continued)

METRIC GAUGE	DIAMETER (MM)	RESISTANCE (OHM/M)
5	0.50	0.0838
6	0.60	0.0582
8	0.80	0.0328
10	1.00	0.0210
14	1.40	0.0107
16	1.60	0.00819
20	2.00	0.00524
25	2.50	0.00335

TABLE C.9 MILLING SPEEDS AND FEEDS (DERIVED FROM TC 9-524)

FEEDS FOR HIGH-SPEED STEEL END MILLS (FEED PER TOOTH IN INCHES)

CUTTER DIAMETER (INCHES)	ALUMINUM	LOW CARBON STEEL	HIGH CARBON STEEL	STAINLESS STEEL
0.125	0.002	0.0005	0.0005	0.0005
0.250	0.002	0.0010	0.0010	0.0010
0.375	0.003	0.0020	0.0020	0.0020
0.500	0.005	0.0025	0.0020	0.0020
0.750	0.006	0.0030	0.0030	0.0030
1.000	0.007	0.0035	0.0030	0.0040
1.500	0.008	0.0040	0.0040	0.0040
2.000	0.009	0.0050	0.0040	0.0050

CUTTING SPEEDS FOR COMMONLY USED MATERIALS/TOOL MATERIAL

WORK MATERIAL	HIGH-SPEED STEEL	UNCOATED CARBIDE	COATED CARBIDE	DIAMOND
Aluminum				
Low silicon	300–800	700–1400		
High silicon				1000-3000
Low carbon steel		60–100	250–350	500–900
Alloy steel	40–70		350–600	

continued on next page

TABLE C.9 MILLING SPEEDS AND FEEDS (DERIVED FROM TC 9-524) (Continued)

WORK MATERIAL	HIGH-SPEED STEEL	UNCOATED CARBIDE	COATED CARBIDE	DIAMOND
Tool steel	40–70		250–500	
Stainless steel				
200 & 300 series		30–80	100–250	400–650
400 & 500 series			250–350	
Nonmetallics		400–600		400–2000
Superalloys		70–100	90–150	

Calculations:

rpm for an endmill uses rpm = CS × 4/D

where CS = Cutting Speed D = Diameter of cutting tool

Feed rate is in inches per minute. ipm = F × N × rpm where F = feed per tooth N = number of teeth

Example:

Calculating feed and speed for a 2-flute 0.5" hss end mill cutting aluminum

rpm = CS × 4/D 300 × 4/.5 = 2400 rpm

ipm = F × N × rpm 0.005 × 2 × 2400 = 24 ipm

TABEL C.10 PLAIN BEARING PMAX, VMAX, AND PVMAX (COPIED FROM MCMASTER CARR)

PLAIN BEARINGS (RADIAL, THRUST AND COMBINATION LOAD)					
BEARING MATERIAL	TEMPERATURE RANGE	FOR SHAFT HARDNESS	PMAX	VMAX	PVMAX
Vespel®	–400 to + 550°F	Hard	4,900	3,000	300,000
Rulon-LR	–400 to +550°F	Hard	1,000	400	10,000
Rulon-J	–400 to +550°F	Soft and up	750	400	7,500
Rulon-641	–400 to +550°F	Soft and up	1,000	400	10,000
Nylon	–20 to +250°F	Medium and up	400	360	3,000
Nylon 642	–40 to +410°F	Soft and up	1,500	400	16,000
MDS-Filled Nylon	–40 to +200°F	Hard	2,000	200	2,700
UHMW	–100 to +180°F	Medium and up	1,000	100	3,000
SAE 841 Bronze	10 to 220°F	Medium and up	2,000	1,200	50,000
SAE 660 Bronze	10 to 450°F	Medium and up	4,000	750	75,000
Graphite	–350 to +800°F	Medium and up	3,500	560	56,000
Graphalloy®	–450 to +750°F	Hard	750	300	12,000

continued on next page

TABEL C.10 PLAIN BEARING PMAX, VMAX, AND PVMAX (COPIED FROM MCMASTER CARR) (continued)

BEARING MATERIAL	TEMPERATURE RANGE	FOR SHAFT HARDNESS	PMAX	VMAX	PVMAX
Teflon/Nomex-Lined Fiberglass	–290 to +320°F	Hard	25,000	150	25,000
Rulon-F-Lined Fiberglass	–290 to +320°F	Soft and up	5,000	400	20,000
Steel-Backed Teflon	–328 to +536°F	Medium and up	36,000	390	51,000
Steel-Backed Teflon Composite	–320 to +300°F	Hard	30,000	600	50,000
Steel-Backed Rulon-LR	–400 to +550°F	Hard	1,500	400	20,000
Steel-Backed Nylon	–40 to +200°F	Hard	750	200	3,500
Aluminum-Backed Frelon	–400 to +500°F	Hard	1,500	140	10,000

TABLE C.11 PROPERTIES OF METALS

MATERIAL	DENSITY (LBS/IN³)	SPECIFIC HEAT (BTU/LB/°F)	THERMAL CONDUCTIVITY BTU / (HR-FT-°F)	MELTING POINT (°F)
Aluminum	0.098	0.24	1488	1220
Brass (Yellow)	0.306	0.096	756	1724
Copper	0.322	0.095	2580	1976
Incoloy 800	0.29	0.13	80	2500
Inconel 600	0.304	0.126	103	2500
Iron, Cast	0.26	0.12	346	2150
Lead, solid	0.41	0.032	216	621
Lead, Liquid	0.387	0.037	108	—
Magnesium	0.063	0.27	1106	1202
Solder (50% Pb-50% Sn)	0.323	0.051	310	361
Steel, mild	0.284	0.122	312	2570
Steel, Stainless 304	0.286	0.012	120	2550

continued on next page

TABLE C.11 PROPERTIES OF METALS (Continued)

MATERIAL	DENSITY (LBS/IN³)	SPECIFIC HEAT (BTU/LB/°F)	THERMAL CONDUCTIVITY BTU / (HR-FT-°F)	MELTING POINT (°F)
Steel, Stainless 430	0.275	0.011	156	2650
Tin, solid	0.263	0.065	408	450
Tin, Liquid	0.253	0.052	218	—
Titanium 99.0%	0.164	0.13	112	3035

TABLE C.12 STOCK ALUMINUM PROPERTIES (DERIVED FROM ENGINEERSEDGE.COM)

ALUMINUM ALLOY INFORMATION			
ALLOY #	ALLOY MATERIAL	DESCRIPTION	TENSILE STRENGTH (PSI)
1100	None (Pure)	Excellent corrosion resistance, workability, and weldability. Not heat treatable.	14,000 to 24,000
5052	Magnesium	Very good corrosion resistance, good workability, weldability. Not heat treatable.	31,000 to 44,000
2024	Copper	Fair workability and corrosion resistance. Heat treatable.	30,000 to 63,000
6061	Magnesium, Silicon	Good formability, weldability, and corrosion resistance. Is heat treatable.	7,000 to 39,000
7075	Zinc, Copper Magnesium and Chromium	Poor formability, good machine ability. Heat treatable.	32,000 to 76,000

TABLE C.13 SHEET METAL GAUGES (COPIED FROM ENGINEERSEDGE.COM)

	NON-FERROUS		STEEL SHEETS		STRIP & TUBING BIRMINGHAM OR STUBS	
GAUGE #	GAUGE DECIMAL	LBS./SQ. FT. 6061 ALUMINUM	GAUGE DECIMAL	LBS./SQ. FT. STEEL STRIP	GAUGE DECIMAL	LBS./SQ. FT. STEEL STRIP
000000	.5800	—	—	—	—	—
00000	.5165	—	—	—	.500	20.40
0000	.4600	—	—	—	.454	18.52
000	.4096	—	—	—	.425	17.34
00	.3648	—	—	—	.380	15.50
0	.3249	—	—	—	.340	13.87
1	.2893	—	—	—	.300	12.24
2	.2576	—	—	—	.284	11.59
3	.2294	—	0.2391	9.754	.259	10.57
4	.2043	—	0.2242	9.146	.238	9.710
5	.1819	—	0.2092	8.534	.220	8.975
6	.1620	2.286	0.1943	7.926	.203	8.281
7	.1443	2.036	0.1793	7.315	.180	7.343
8	.1285	1.813	0.1644	6.707	.165	6.731
9	.1144	1.614	0.1495	6.099	.148	6.038
10	.1019	1.438	0.1345	5.487	.134	5.467
11	.0907	1.280	0.1196	4.879	.120	4.895
12	.0808	1.140	0.1046	4.267	.109	4.447
13	.0720	1.016	0.0897	3.659	.095	3.876
14	.0641	.905	0.0747	3.047	.083	3.386
15	.0571	.806	0.0673	2.746	.072	2.937
16	.0508	.717	0.0598	2.44	.065	2.652
17	.0453	.639	0.0538	2.195	.058	2.366
18	.0403	.569	0.0478	1.95	.049	1.999
19	.0359	.507	0.0418	1.705	.042	1.713
20	.0320	.452	0.0359	1.465	.035	1.428
21	.0285	.402	0.0329	1.342	.032	1.305
22	.0253	.357	0.0299	1.22	.028	1.142
23	.0226	.319	0.0269	1.097	.025	1.020
24	.0201	.284	0.0239	0.975	.022	.898
25	.0179	.253	0.0209	0.853	.020	.816
26	.0159	.224	0.0179	0.73	.018	.734
27	.0142	.200	0.0164	0.669		

TABLE C.14 PROPER GEAR USAGE (DERIVED FROM ENGINEERSEDGE.COM)

TYPE	PRECISION RATING	COMMENTS ON PRECISION
Spur	Excellent	Parallel shafts. High speed, load, and efficiency. Recommended for all gear meshes except where right-angle drives cannot be avoided.
Bevel	Fair to good	Intersecting shafts. High speed and load. Recommended for right-angle drive, particularly low ratios. Should be located at one of the less critical meshes of the drivetrain.
Helical	Good	Parallel shafts. High speed and load. Efficiency slightly less than spur. Equivalent quality to spurs except for complication of their angle. Recommended for all high-speed and high-load meshes. Axial thrust component must be accommodated.
Worm	Fair to good	Right-angle shafts. High speed and load. Low efficiency. Recommended for combination high speed and right-angle drive. High sliding requires lubrication.

TABLE C.15 MOTOR SPEC CHART (PART 1) (COURTESY OF JOHN HOFFMAN)

MOTOR	MODEL NUMBER	NOM VOLTS	MAX VOLTS	MAX RPM	NO LOAD CURRENT (AMPS)	STALL CURRENT (AMPS)	STALL TORQUE (IN-LB)	KT IN-OZ/ AMP	KV RPM/VOLT
Bosch	GPA 750	24	36	4,000	3.0	180	98.0	8.7	167
EV Warrior	GPB?	24	24	5,000	3.0	120	50.0	6.7	208
Jensen	RCM-140	12	24	3,000	2.0	80	27.0	5.4	250
Mag Motor	C40-ZP-300FX	24	36	4,000	3.3	460	233.0	8.1	167
Scott	4BB-02488	24	36	3,300	5.5	440	270.0	9.8	138
Leemco	LEM-200	24	60	3,000	1.0	320	250.0	12.5	125
Leemco	WEV-200	50	?	3,500	1.0	160	190.0	19.0	70
NPC 1200	1200	24	36	3,000					125
NPC 02423	2423	24	36	3,300		34	25.0	11.8	138
NPC 01280	1280	12	24	2,000		50	30.0	9.6	167
NPC 60522	60522	24	36	120					5
NPC 02486	2486	24	36	2,000					83
Cobalt 640	640	18	26	12,500	2.0	198	17.0	1.4	694
DeWalt	18v Low Gear	18	24	450		110	400	21.8	89
DeWalt	24v	24	36	450		50	550.0	176.0	19
Aveox	1415/4Y	24	36	30,000		70	16.0	2.3	595
Aveox	2315 /6Y	24	36	10,000				3.8	360
Ecycle	MG 24	120	160	6,000		70	150.0	34.3	39
Ecycle	MG 13	12	48	3,600		100	110.0	17.6	78

TABLE C.15 MOTOR SPEC CHART (PART 2) (COURTESY OF JOHN HOFFMAN

MOTOR	NOM HP	PEAK HP	WT (LB)	EFF.	$/HP	LB/HP	SOURCE**	COST**
Bosch		1.45	8.5	75%	$119	5.9	Team Delta	$172
EV Warrior	0.25	0.8	3.0	75?	$19	3.8	MECI	$15
Jensen		0.9	3.0	low?	$73	3.3		$65.50
Mag Motor		3.5	11.9	84%	$99	3.4	Robot Books	$345
Scott	1.00	3.5	17.0	79%	$74	4.9	Wilde	$260
Leemco	9.00	12.6	25.0	90%	$91	2.0	Wilde	$1,145
Leemco		9.5	21.5	85%	$55	2.3	Wilde	$525
NPC 1200	1.50	2	9.7		$63	4.9	NPC	$125
NPC 02423		0.35	2.0		$180	5.7	NPC	$63
NPC 01280		0.3	2.0		$217	6.7	NPC	$65
NPC 60522		1.14	15.0	60%	$162	13.2	NPC	$185
NPC 02486	0.25	0.6	8.0		$183	13.3	NPC	$110
Cobalt 640	0.75	1.2	0.8	80%	$125	0.7	Astroflight	$150
DeWalt		1.2	1.6		$50	1.3	DeWalt	$60
DeWalt	0.25	1	2.8	60%	$85	2.8	DeWalt	$85
Aveox	0.90	1.8	0.8	90?	$138	0.4	Aveox	$248
Aveox		4	1.9	90?	$220	0.5	Aveox	$880
Ecycle		12.6	13.2	94%	$23	1.0	Acycle	$295
Ecycle	5.60	6.4	13.2	94%	$46	2.1	Ecycle	$295

** At the time of printing.

TABLE C.16 BOLT GRADE, TENSILE STRENGTH, TORQUE SPECIFICATION, AND SHEAR STRENGTH***

SAE GRADE (# OF MARKS)	DIA. (IN)	COARSE THREAD (POUNDS)	FINE THREAD (POUNDS)	MINIMUM YIELD STRENGTH (PSI)	MINIMUM TENSILE STRENGTH (PSI)	NOMINAL ASSEMBLY TORQUE (FT-LBS)**	
						COURSE THREAD	FINE THREAD
2	1/4	2350	2700	36,000*	60,000*	5.5	6.3
No Marks	5/16	3900	4300			11	12
	7/16	7850	8800			32	36
	1/2	10500	11800			50	55
	9/16	13500	15000			70	80
	5/8	16700	18900			100	110
	3/4	24700	27600			175	200
5	1/4	3800	4350	92,000	120,000	8	10
Three	5/16	6300	6950	thru	thru	17	19
Marks	7/16	12800	14400	81,000	105,000	50	55
	1/2	17000	19200			75	85
	9/16	21800	24400			110	120
	5/8	27100	30700			150	170
	3/4	40100	44800			260	300
8	1/4	4750	5450	130,000	150,000	12	14
Six Marks	5/16	7850	8700			24	27
	7/16	15900	17800			70	80
	1/2	21300	24000			110	120
	9/16	27300	30400			150	170
	5/8	33900	38400			210	240
	3/4	50100	56000			380	420

*** Some bolts have the same head identification but different strengths.

** Do not tighten past the crushing point of the materials you are holding together.

* You cannot count on grade 2 bolts actually having these minimums.

TABLE C.17 OUTSIDE DIAMETER, NUMBER OF TEETH, AND PITCH OF STANDARD SPROCKETS (COPIED FROM MCMASTER CARR)

	OUTSIDE DIAMETER (OD) OF THE SPROCKET					
NO. OF TEETH	#25 1/4" PITCH	#35 3/8" PITCH	#40 AND #41 1/2" PITCH	#50 5/8" PITCH	#60 3/4" PITCH	#80 1" PITCH
9	0.84"	1.26"	1.67"	2.09"	2.51"	3.35"
10	0.92"	1.38"	1.84"	2.30"	2.76"	3.68"
11	1.00"	1.50"	2.00"	2.50"	3.00"	4.01"
12	1.08"	1.63"	2.17"	2.71"	3.25"	4.33"
13	1.17"	1.75"	2.33"	2.91"	3.49"	4.66"
14	1.25"	1.87"	2.49"	3.11"	3.74"	4.98"
15	1.33"	1.99"	2.65"	3.32"	3.98"	5.30"
16	1.41"	2.11"	2.81"	3.52"	4.22"	5.63"
17	1.49"	2.23"	2.98"	3.72"	4.46"	5.95"
18	1.57"	2.35"	3.14"	3.92"	4.70"	6.27"
19	1.65"	2.47"	3.30"	4.12"	4.95"	6.59"
20	1.73"	2.59"	3.46"	4.32"	5.19"	6.91"
21	1.81"	2.71"	3.62"	4.52"	5.43"	7.24"
22	1.89"	2.83"	3.78"	4.72"	5.67"	7.56"
23	1.97"	2.95"	3.94"	4.92"	5.91"	7.88"
24	2.05"	3.07"	4.10"	5.12"	6.15"	8.20"
25	2.13"	3.19"	4.26"	5.32"	6.39"	8.52"
26	—	3.31"	4.42"	5.52"	6.63"	8.84"
28	—	3.55"	4.74"	5.92"	7.11"	9.48"
30	2.53"	3.79"	5.06"	6.32"	7.59"	10.11"
32	—	4.03"	5.38"	6.72"	8.07"	10.75"
34	—	—	5.70"	7.12"	8.54"	11.39"
35	2.93"	4.39"	5.86"	7.32"	8.78"	11.71"
36	—	4.51"	6.02"	7.52"	9.02"	12.03"
38	—	—	6.33"	—	9.50"	12.67"
40	3.33"	4.99"	6.65"	8.32"	9.98"	—
42	—	5.23"	6.97"	8.72"	10.46"	13.94"
45	3.73"	5.59"	7.45"	9.31"	11.18"	—
48	—	5.95"	7.93"	9.91"	11.89"	15.86"
60	—	7.38"	9.84"	12.30"	14.76"	19.68"

TABLE C.18 DIMENSIONS OF SQUARE-END MACHINE KEYS

	SQUARE KEYS	
SHAFT DIAMETER (INCHES)	WIDTH AND THICKNESS (INCHES)	BOTTOM OF KEYWAY TO OPPOSITE SIDE OF SHAFT (INCHES)
1/2	1/8	0.430
9/16	1/8	0.493
5/8	3/16	0.517
11/16	3/16	0.581
3/4	3/16	0.644
13/16	3/16	0.708
7/8	3/16	0.771
15/16	1/4	0.796
1	1/4	0.859
1 – 1/16	1/4	0.923
1 – 1/8	1/4	0.996
1 – 3/16	1/4	1.049
1 – 1/4	1/4	1.112
1 – 5/16	5/16	1.137
1 – 3/8	5/16	1.201
1 – 7/16	3/8	1.225
1 – 1/5	3/8	1.289
1 – 9/16	3/8	1.352
1 – 5/8	3/8	1.416
1 – 11/16	3/8	1.479
1 – 3/4	3/8	1.542
1 – 13/16	1/2	1.527
1 – 7/8	1/2	1.591
1 – 15/16	1/2	1.655
2	1/2	1.718

TABLE C.19 WEIGHTS OF MATERIALS

MATERIAL	WEIGHT (LBS/FT^3)	WEIGHT (LBS/IN^3)
Metals		
Aluminum	168.5	0.0975
Brass, 80% C, 20% Z	536.6	0.3105
Brass, 70% C, 30% Z	526.7	0.3048
Brass, 60% C, 40% Z	521.7	0.3019
Brass, 50% C, 50% Z	511.7	0.2961
Copper	559.5	0.3210
Gold	1206.3	0.6969
Iron, Cast	450	0.2604
Iron, Wrought	486.7	0.2817
Lead	707.7	0.4095
Magnesium	108.6	0.0628
Molybdenum	636.5	0.3683
Nickel	549	0.3177
Platinum	1333.5	0.7717
Silver	657	0.3802
Steel	490	0.2836
Titanium	280.1	0.1621
Tin	455	0.2632
Tungsten	1205.1	0.6974
Plastics		
Polycarbonate	76.2	0.0441
Nylon	70.6	0.0408
Polyethylene	59.3	0.0343
Polypropylene	56.2	0.0325
Other		
Brick, Common	112	0.0648
Concrete	137	0.0793
Earth, Common	75	0.0434
Earth, Packed	100	0.0579
Glass	162	0.0938
Ice	56	0.0324
Petroleum Gasoline	42	0.0243
Water, Fresh	62.5	0.0362
Water, Sea	64	0.0370

WEB SITES

McMaster Carr (mechanical stuff)
http://www.mcmaster.com

How Stuff Works (Force, Power, Torque and Energy Work)
http://www.howstuffworks.com/

Index of Mechanism Terminologies (Gears)
ttp://www-2.cs.cmu.edu/~rapidproto/mechanisms/chpt2.html#HDR31

Martin Sprocket & Gear Inc. (Sprockets, Belts, and Gears)
http://www.martinsprocket.com/home.htm

Stock Drive Products / Sterling Instrument Technical Library (Mechanical)
http://www.sdp-si.com/Sdptech_lib.htm

Engineers Edge (lots of charts)
http://www.engineersedge.com

Supersite of Reade Advanced Materials (charts)
http://www.reade.com/Conversion/wire_gauge.html

Team Slam (gas engine info)
http://users.intercomm.com/stevenn/slamweb/Slam1.htm

Eric Behr's Collection of Electronic Speed Controls (speed controller stuff)
http://www.math.niu.edu/~behr/RC/speed-ctl.html

World and Regional Paintball Information Guide (Pneumatics)
http://www.warpig.com/

Cadex (battery stuff)
http://www.cadex.com/b_02_3_1_charging.asp

Panasonic Sealed Lead Acid Batteries
http://www.panasonic.com/industrial/battery/oem/chem/seal/index.html

Magellan Technologies (Sub C NiCds)
ttp://www.magtechinc.net/subc.htm

Hawker Energy Products Inc. (Hawker battery info)
http://www.hepi.com/genesis.htm

MatWeb Online Materials Information Resource
http://www.matweb.com/

Monroe Engineering Stuff (drills and taps chart information)
http://www.monroeengineering.com/helpfulstuff/

Headquarters Department of the Army
(TC 9-524 Fundamentals of Machine Tools)
http://155.217.58.58/cgi-bin/atdl.dll/tc/9-524/toc.htm

Allsup Consulting (glossary terms)
http://www.dhc.net/~allsup/hot.htm

Team Saber Competitive Robotics (Glossary terms and design hints)
http://www.teamsaber.com/

Roberts Gadgets and Gismos Technology Data and References
http://www.bpesolutions.com/gadgets.ws/gtechgen.html

RBE Electronics (wire size chart)
http://www.rbeelectronics.com/wtable.htm

Pain Enterprises, Inc. (CO_2 info)
http://www.painenterprises.com/dryicemsds.html

Carbon Dioxide Review, edited by W.C. Clark, Oxford University Press, New York, 1982, p. 468.
http://cdiac.esd.ornl.gov/pns/convert.html

Motor specs chart provided by John Hoffman.

FURTHER READING

ELECTRONICS

Teach Yourself Electricity and Electronics, Stan Gibilisco
ISBN: 0071377301

The Basics of Soldering, Rahn
ISBN: 0471584711

Modern Electronics Soldering Techniques, Singmin
ISBN: 0790611996

The Forrest Mims Engineer's Notebook, Mims
ISBN: 1878707035

The Art of Electronics, Horowitz & Hill
ISBN: 0521370957

Mechanical Devices for Electronics Experimenters, Rorabaugh
ISBN: 0070535477

Electronic Devices, Floyd
ISBN: 013028484X

Easy PIC'n, Benson
ISBN: 0965416208

Design With PIC Microcontrollers, Peatman
ISBN: 0137592590

The Embedded PC's ISA Bus : Firmware, Gadgets, and Practical Tricks, Nisley
ISBN: 157398017X

Handbook of Batteries, Linden
ISBN: 0070379211

Electric Motors and Control Techniques, Gottlieb
ISBN: 0070240124

Astroflight Inc., Electric Motor Handbook, Boucher
ISBN: 0964406500

MACHINING

Basic Machining Reference Handbook, Meyers & Slattery
ISBN: 0831131209

Machinery's Handbook 26, Industrial Press
ISBN: 0831125756

Tabletop Machining, Martin
ISBN: 0966543300

An Introduction to CNC Machining and Programming, Gibbs & Crandell
ISBN: 0831130091

Basic Jig and Fixture Making for Metalworking Trainees, Harig
ISBN: 0910399107

Fundamentals of Machining and Machine Tools, Boothroyd & Knight
ISBN: 0824778529

How to Read Shop Prints and Drawings with Blueprints, Hardman
ISBN: 0910399018

Machining Fundamentals, Walker
ISBN: 1566376629

Measuring and Gaging in the Machine Shop, National Tooling and Machining Assn. Staff
ISBN: 0910399271

ENGINEERING

The Science and Design of Engineering Materials, Schaffer
ISBN: 0072448091

Engineer To Win, Smith
ISBN: 0879381868

Nuts, Bolts, Fasteners, and Plumbing Handbook, Smith
ISBN: 0879384069

High Performance Hardware, Aird
ISBN: 1557883041

Newnes Engineering Materials Pocket Book, Bolton
ISBN: 0750649747

The Robot Builder's Bonanza 2nd Edition, McComb
ISBN: 0071362967

Mechanisms and Mechanical Devices Sourcebook 3rd Edition, Sclater and Chironis
ISBN: 0071361693

Illustrated Sourcebook of Mechanical Components, Parmley
ISBN: 0070486174

Millwrights and Mechanics Guide, Nelson
ISBN: 002588591X

Ingenious Mechanisms for Designers and Inventors, Horton & Gartner
ISBN: 0831110848

Five Hundred and Seven Mechanical Movements, Brown
ISBN: 1879335638

The Complete Guide to Chain, U.S. Tsubaki, Inc.

OTHER

ARC Welding, Walker
ISBN: 1566374774

Welder's Handbook, Finch
ISBN: 1557882649

The Sponsorship Seeker's Toolkit, Grey & Skildum-Reid
ISBN: 0074707078

The Technology of Fluid Power, Reeves
ISBN: 0827366647

Sheet Metal Handbook, Fournier
ISBN: 0895867575

Racer's Encyclopedia of Metals, Fibers & Materials, Aird
ISBN: 0879389168

PARTS CATALOGS

(Look them up on the Internet and ask for them.)

- Aircraft Spruce & Speciality Co.
- Allied Electronics
- Digi-Key
- Gates Industrial Power Transmission Products (Synchronous belts)
- Gates PowerGrip GT2 Belt Drive Design Manual
- Grainger Industrial Supply
- Harbor Freight
- Lab Safety Supply (Safety and Material Handling catalogs)
- Martin Sprocket & Gear (Excellent products and web catalog)
- McMaster Carr Supply Company (Good luck. Check web catalog.)
- MSC Industrial Supply Company
- Newark Electronics

- Small Parts Inc.
- Stock Drive Products (Get them all)
- W.M. Berg, Inc.
- Wick's Aircraft Supply
- Tower Hobbies

FREQUENTLY ASKED QUESTIONS

General Questions*

Q. How do I convert between ounce-inches (oz-in) and pound-feet (lb-ft)?

A. To convert lb-ft to lb-in, multiply by 12.
To convert lb-in to oz-in, multiply by 16.
1 lb-ft = 12 lb-in = 192 oz-in.
lb-ft = oz-in divided by 192.

Q. How do I find motor horsepower if I have the current and voltage at which the motor runs?

A. Find the power in watts and divide by 746:
Power (watts) = Current (amps) × Voltage (volts)
Power (hp) = Power (watts) / 746

DC motors are not usually more than 80 percent efficient. Some are worse. Factor that into your Power (watts) calculation by reducing the watts by 20 percent or more.

ST = torque at motor at stall* WT = torque at the wheel* AT = torque at the axle*
R = radius of the wheel in inches D = diameter of the wheel in inches
r = gear ratio RPM = revolutions per minute of the wheel
* = all torque measured in pound-inches (lb-in) unless otherwise noted.

Q. I have a motor with 47 pound-feet (lb-ft) of stall torque. I want two of them to drive a 100-pound bot with 8-inch wheels. What gear ratio will I need? They spin at 3000 rpm, how fast will it go in miles per hour?

A. We need a total of at least 50 pound-feet from each motor for the bot to carry its own weight. Wait! That's not good planning! Plan to carry your opponent. That means we need at least 100 pound-feet from each motor. Gear/sprocket/pulley reductions multiply torque by their ratios. You have to find the ratio (r) that, when multiplied by 24 in-lbs at stall (ST), gives 100 in-lb at the wheel (WT). Don't forget that the radius (R) of the wheel decreases torque. So you have to divide by R and overcome that decrease:

$WT = (ST \times r) / R$

Solve for r:

$r = (WT \times R) / ST$
$r = (100 \times 4) / 47$
$r = 400 / 47$
$r = 8.5$

So you need an 8.5 to 1 ratio to drive this bot. Remember that you shouldn't go above about a 5 to 1 ratio. This means you should use a two-stage reduction.

As for the second part of the question, remember that gear/sprocket/pulley reductions divide speed by their ratios. Given a ratio (r), a wheel diameter (D), and an rpm of the motors ,you can find the speed in mph.

Wheel rpm is the motor rpm divided by the gear ratio:

$rpm = (Motor\ rpm) / r$
$rpm = 3000 / 8.5$
$rpm = 352.9$

Figuring out mph:

$mph = (rpm \times D \times 188.4) / 63360$
$mph = (352.9 \times 8 \times 188.4) / 63360$
$mph = (2823.2 \times 188.4) / 63360$
$mph = 531890.88 / 63360$
$mph = 8.4$

While 8.4 mph is not bad, it is not fast either. You can sacrifice push power (torque) for more speed if you wish. Simply change the gear reduction to something smaller.

Q. 8.4 mph is too slow for my taste. How do I make it go 12 mph?

A. You are really going to do all the previous calculations over but in reverse order, so that you can make sure that with the new gear reduction, your bot will be able to at least carry it's own weight.

Given the 12 mph speed and 8-inch wheels, you can find the rpm of the wheel:

$rpm = (63360 \times mph) / (188.4 \times D)$
$rpm = (63360 \times 12) / (188.4 \times 8)$
$rpm = 760320 / 1507.2$
$rpm = 504.46$

Given 3,000 rpm of the motor without gears and the 504.46 rpm of the wheel with gears, you can find the gear reduction ratio:

r = (rpm of the motor) / (rpm of the wheel)
r = 3000 / 504.46
r = 5.95

Now that you have the gear ratio required to go 12 mph with these motors, you can find the amount of torque the wheels will have at stall:

$WT = (ST \times r) / R$
$WT = (47 \times 5.95) / 4$
$WT = 279.65 / 4$
$WT = 69.91$

This 69.91 lb-in is for one wheel. Multiply that by the two motors you have and the total torque for your bot is 139.82. This is more than enough to carry your bot's own weight. It is not enough to carry your opponent. I would shoot for a lower top speed and higher carrying capacity.

Q. Is this motor a good motor for my bot?

A. The easiest answer in the world is to say "Buy it, build it, and see for yourself." However, since this is a book about building these robots, I'll give at least a partial answer. Look at the Motor Spec Charts part one and two in Appendix C. If the motor you are asking about isn't in there, check the specs for your motor against the specs for any of the motors on the chart. If they are similar, your motor *might* be a good one to use in your design.

Q. How do I attach a gear, sprocket, or pulley to my motor shaft?

A. This is covered in more detail in the Electric Motors section and the Axles, Bearings and Bushings section of this book. Usually people attach a gear, sprocket, or pulley to their motor shaft in one or more of a few different ways. Some motor shafts have a keyway cut into them. In this case, the gear/sprocket/pulley should have the same type of keyway in it. Key stock is inserted and held in place by a set screw. Some people use only a set screw to hold the part to the motor shaft. This is a bad idea. Some people "press" the part onto the motor shaft. This is accomplished by boring out the part to tight tolerances so that you can put the two together by hand. Once on the shaft, the part won't come off except under extreme loads. Other people use square or octagonal shafts and bores. This is difficult for even the most advanced hobbyist.

Q. How do I attach my wheels to my drive shaft or drill?

A. See the answer for the previous question. However, there are a couple differences when it comes to wheels. Some wheels require special hubs that make mounting them on standard size shafts easier. Check with the Web sites that sell wheels. They usually sell hubs along with them.

Q. Do EV Warriors spin in both directions?

A. The popular EV Warrior motors do spin in both directions. If you've been doing enough research to ask this question, it isn't surprising that you got confused about the motors. EV Warrior motors are "timed" to rotate faster and stronger in one direction or the other depending on how it is labeled. If it is labeled as "CCW" (counter clockwise) then it will rotate faster and stronger in the CCW direction. If it is labeled as "CW" (clockwise), the opposite is true. When running the motor, you can audibly hear the difference in speed. There are people who have experimented with neutral timing on these motors with some success. This involves modifying the brush housing, so I do not recommend attempting the operation while you are still new to the sport. In other words, you have enough to worry about without that extra project.

Q. How can I mount my drill motor?

A. Drill motors are popular because they have a gearbox on them already and are fairly strong for their small size. Mounting them can be a pain though. Some people have gone the simple route and left the motor and gearing inside the drill housing. They run the wires outside the housing and connect them to the speed controller. Then they use strapping, hose clamps, or some other method to hold the entire drill housing down on the base plate of the robot. Some people go further and machine blocks of aluminum or plastic so that they fit around the motor and gear housing. They secure the motor and housing to the aluminum or plastic blocks in some manner and mount the entire thing on the base of the robot. Some people enclose the entire motor and gear housing inside a suitable material and mount that assembly to the base plate. How you do it is entirely up to your mechanical ability and/or wallet. You may be able to get some good ideas if you go to your local machine shop and ask their opinion. They may even build it for you.

Q. Can I use a Rebel/Rooster/RC-type speed controller in my bot?

A. I included the most common types of speed controllers in Chapter 4 of the book. I left the small, RC-type speed controllers out because they aren't usually suitable for the robots we are building. A large number of builders are constructing bots that weigh 1, 12, and 30 pounds. These small speed controllers may work nicely in the lighter weight bots. There are some drawbacks that are controller specific. Some controllers may have a reverse delay. This means they have a delay between the time they are going full speed forward and switched to full speed reverse. It is built in so that voltage and current spikes don't fry the controller when switched on and off rapidly. Also, many of these controllers do not even run the motors in reverse. If you decide to use one of these controllers, use it in a small robot and make sure it can handle the current that your motors demand.

Q. What does it mean to "properly size a speed controller to my motor"?

A. Speed controllers are generally electronic devices. Electronic devices are made to handle either an exact voltage and exact current or a small range of voltages and currents. Many electronic devices, RC car speed controllers for example, are not made to handle the amount of current that is required by motors that are strong enough for full size battling robots. If you used a 1-horsepower wheelchair motor along with a Tekin Rebel RC car speed controller, the controller would soon be overwhelmed by the amount of current it would try to deliver to the motor. The smoke would roll instead of the bot. Read the chapters on speed controllers, batteries, and electric motors to find out how to decide which speed controller will safely control your motors.

Q. Can I use a windshield wiper motor or starter motor in my bot?

A. Yes, you can, if it has enough torque to carry the robot at the speed you want. There are drawbacks to using windshield wiper motors and starter motors. Both types of motors were designed to be used intermittently in cars. Neither motor is likely to stand up to constant use in the conditions that exist during a robot fight. Neither motor is terribly efficient either. Cars supply power to their electrical systems with the alternator. The battery is not in constant use. In fact, the battery is usually in a "charging" state. The alternator is capable of sending all the current required by either motor and then some. Because it is pretty much a bottomless pit of energy, there was no reason to design the motors to be efficient. The batteries in your robot are not bottomless pits. They die in a hurry and you will do better to find a more efficient motor. Some starter motors do not run in reverse; some others require special speed controllers to make them run. With the selection of surplus motors in other markets, starter motors usually are not worth the trouble.

Q. Can I use two speed controllers on one motor so that the combined current capability can handle the stronger motor?

A. No, is the short answer. You cannot wire two separate, electronic speed controllers so that they share the current of a motor that draws more amps than a single controller's rating. However, there are speed controllers that can be linked to do the job. This feature is usually only enabled by the factory and cannot be enabled by the customer. Also, there are some motors that have four, electrically separate brushes and separated windings. With these motors it is possible to connect one speed controller to one pair of brushes, another controller to the other pair of brushes, and command the two speed controllers with the same signal. This has been done, but remember that you do this at the risk of frying some expensive equipment. Usually, when you skimp on something to save money, you end up spending more than you would have if you just bought the right part in the first place.

Q. I've heard that by using gear reduction, I can keep from burning up my speed controllers. Is this so?

A. Sort of. Speed controllers only supply the amount of current for which they are rated. More current than that will fry them no matter what. Some controllers use *current limiting* to help protect against that. Gear reduction decreases motor speed and increases torque applied to the ground through the wheel. Gear reduction will only make it more difficult to stall the drive motors, therefore keeping the current below the speed controller rating. However, it is not possible to guarantee against stalling the motor.

Q. How do I figure out if my gear reduction and motor setup will fry the speed controllers or help them live longer? My bot weighs 120 pounds with four, 10-inch diameter wheels. It has one motor per wheel. I am using the Bosch GPA 750 motor. I also have a 10:1 gear ratio.

A. If you have not read it yet, check out the "Finding Current Draw" section of Chapter 6. In it I give formulas for answering this question and explain a little about why they work. However, there are three basic steps to answering this question:

1. Figure out the estimated weight supported by one wheel.
2. Convert that into torque at the motor using the wheel radius and gear ratio.
3. Convert that into amps at the motor using the torque constant (Kt) for your motor.

If you have a 120-pound robot with four wheels (one motor per wheel) and can assume that each wheel supports an equal amount of weight, the estimated weight supported by one wheel will be about 30 pounds. In formula form that is:

Sw = Supported weight on one wheel
W = Total bot weight

N = Number of wheels on ground
Sw = W / N
Sw = 120 / 4
Sw = 30 pounds

Multiply that by 16 to convert the answer to ounces (oz). That gives you 480 oz. Keep the ounce units because the Kt is usually specified in ounce-inches per amp (oz-in/A):

Sw = Sw × 16
Sw = 30 × 16
Sw = 480 ounces

That number is also the amount of torque your motor produces through the gear train, where the wheel meets the ground. You need to figure out the actual torque produced at the motor before the gear train amplifies it. Do that by multiplying Sw by the wheel radius (R) and then dividing the answer by the gear ratio:

T = Motor torque
R = Wheel radius
r = Gear ratio
T = (Sw × R) / r
T = (480 ounces × 5) / 10
T = 2400 / 10
T = 240 oz-in

Now, convert the torque at the motor (T) into amps (A) drawn by the motor using the torque constant (Kt). Use the "Motor Spec Chart" in Appendix C to find the Kt of the Bosch GPA 750. If your motor is not in the chart or you do not have the Kt available from the manufacturer, you simply divide the measured stall torque in ounce-inches by the measured stall current in amps. Be careful: the "Motor Specs Chart" lists stall torque in pound-inches (lb-in). Convert it to inch-ounces (in-oz). In this case, the Bosch GPA 750 Kt is 8.7 in-oz per amp:

A = T / Kt
A = 240 / 8.7
A = 27.6 amps

Your motors will only draw about 28 amps when the wheels begin to spin in this configuration. There are many variables that will change this value. Your opponent could lift one side so that your bot is riding on only two wheels. He could lift you so that you ride on only one wheel. He could get on top of you and double your weight. Your center of gravity could possibly not be the center of the bot. The CG definitely changes if your opponent lifts you or gets on top. You can never calculate all the correct answers, but this one will give you a ballpark idea. By the way, you are using too many motors to drive this 120-pound bot. Use the weight for more armor and/or a stronger weapon.

Battery Questions*

Q. With two drive motors that draw 40 stall amps each, how much battery power do I need for a 5-minute match?

A. You obviously will not stall your motors for an entire 5-minute match, but if you figure out the problem based on stall current, you will always have plenty of battery power. If you are tight on weight, you can estimate the average current that the motors will draw over the entire 5 minutes and use that as a basis for your calculations. If you install current sensors and recording capabilities in your bot, you can make several 5-minute test runs and find the true average current draw. Otherwise, we'll do our calculations based on the stall current. In any case, substitute your total current draw for "I" in the equations:

For an SLA battery	For a NiCd or NiMh battery
AH = ((I × T) / 60) × 1.6)	AH = ((I × T) / 60) × 1.15)
AH = ((80 × 5) / 60) × 1.6)	AH = ((80 × 5) / 60) × 1.15)
AH = (400 / 60) × 1.6)	AH = (400 / 60) × 1.15)
AH = 6.67 × 1.6	AH = 6.67 × 1.15
AH = 10.67	AH = 7.67

Notice the difference between SLA and NiCd/NiMh ratings? That's because of the efficiency of each type. It's up to you to decide whether or not the weight savings of the NiCd/NiMh outweighs the cost savings of the SLA. Also, the ratings can be lowered a bit because we are not going to run at stall current the entire time. Once you get a general number, you can use trial and error to get it as close as you need.

Q. Which is better, SLA, NiCd, or NiMh?

A. This depends on the design of your bot and the fatness of your wallet. The major differences between the types of batteries are outlined in the Batteries section of the book. However, it is a fact that NiCd and NiMh batteries are more expensive and require more expensive chargers. For example, a 12-volt, 13-Ah, Hawker SLA will cost, on the upper end, about $150. A charger that can bring it up to fighting potential will run about $40, for a total of $190. You need 10, $3 cells to build a 12-volt, 3-Ah, NiCd battery pack. You need 4 of these packs to get only 12 Ah. You say that's only $120. I say don't forget that you can only get about 30 amps continuous from this pack. If your motors require 90 amps, you'll need 2 more packs for a total of $360. The SLA will supply the 90 amps by itself. This doesn't include the labor cost of building the packs. Plan on putting at least $200 in for a charger and power supply for the NiCds. NiCds have their advantages, too. Read the earlier section of the book and make your own decision on which battery type is better for your robot.

Q. How do I charge my batteries?

A. Standard SLAs charge differently than Hawker/Odyssey SLAs. NiCd and NiMh batteries charge differently than either of the types of SLA batteries. Read the batteries section of this book to find out how it's done. One thing I will say is—do not connect a charger to a battery until you know it will work for that battery. You can destroy an expensive part of your bot if you charge batteries incorrectly.

I = Total Stall Current* T = Match time limit* AH = Amp-hour rating*

Q. How long will it take to charge a 12-volt, 16-amphour battery if it can handle a 15-amp charge?

A. You need to put between 1.05 and 1.10 times the amphour rating back into a drained battery. Do the math and you see that you need to put 17.6, back into the battery. 17.6 amphours divided by 15 amps equals 1.17 hours. Keep in mind that most batteries will not handle a 15-amp charge cycle. Most batteries require 0.4 CA or a 6.4-amp charge for a 16-amphour battery. 17.6 amphours divided by 6.4 amps equals 2.75 hours to charge a normal battery.

Q. Some batteries are listed with a CA 5 Sec, and a CCA rating. What are those?

A. Batteries that are used in cars, boats, lawnmowers, motorcycles, or other vehicles are rated in cranking amps (CA) and/or cold cranking amps (CCA). This is because the battery is only really used heavily while it's trying to start the engine in a vehicle. The manufacturers label the battery ratings in the manner which is most useful to the prospective customer. There are Web sites and catalogs that give both the cranking ratings and the amphour ratings for batteries. For building bots, you will need to use the amphour ratings.

Speed Conversion Questions*

Q. I have an 8-inch diameter wheel and a gear motor that runs at 600 rpm. How many miles per hour will it go? What about feet per second?

A. A good answer dictates that I mention that you don't go anywhere if the motor torque isn't high enough to carry the weight of the bot. The simple answer is 14.27 mph and 20.93 fps. Here is how to find it:

Find mph given rpm and D:
mph = (rpm × D × 188.4) / 63360
mph = (600 × 8 × 188.) / 63360
mph = (4800 × 188.4) / 63360
mph = 904320 / 63360
mph = 14.27

Find fps given rpm and D:
fps = (rpm × D × 3.14) / 720
fps = (600 × 8 × 3.14) / 720
fps = (4800 × 3.14) / 720
fps = 15072 / 720
fps = 20.93

Q. My bot runs 12 miles per hour. What's that in feet per second? My other bot runs at 14.6 feet per second. What's that in miles per hour?

A. Find fps given mph:
fps = (mph × 5280) / 3600
fps = (12 × 5280) / 3600
fps = 63360 / 3600
fps = 17.6

Find mph given fps:
mph = (fps × 3600) / 5280
mph = (14.6 × 3600) / 5280
mph = 52560 / 5280
mph = 9.95

rpm = Revolutions per minute* mph = Miles per hour* fps = Feet per second*
D = Diameter of the wheel in inches*

Q. I want to go 15 miles per hour, and I've got a gear motor strong enough to carry my bot. It turns at 600 rpm. What diameter wheel do I need to get that speed?

A. Find D given fpm and mph:

D = (63360 × mph) / (188.4 × rpm)
D = (63360 × 15) / (188.4 × 600)
D = 950400 / 113040
D = 8.4

Q. I want to go 15 miles per hour, and I've got a 12-inch diameter wheel. What rpm does it need to turn?

A. Find rpm given D and mph:
rpm = (63360 × mph) / (188.4 × D)
rpm = (63360 × 15) / (188.4 × 12)
rpm = 950400 / 2260.8
rpm = 420.4

Pneumatics Questions

Q. Is it legal for a bot to be propelled by pneumatics?

A. Yes, it is legal in the competitions that I know about. However, why would you want to do that? Motors that are powered by air require a lot of it. More air than you would be able to store on the robot. Given that, you would need some type of air compressor on board. Small, battery-powered compressors do not supply enough pressure at a good enough flow rate to gain any advantage over running electric motors directly off the batteries. Gas-powered compressors might do the trick, but you get a lot more power if you just use the engine to drive the wheels. Then, of course, there is the amount of weight you use up when installing the air tank and other pneumatic goodies.

Q. I want to use a (CO_2 or HPA) tank to power a cylinder on my weapon. What size cylinder should I use?

A. The only definite answer I can give you is … bigger is better. You have to balance the size of your cylinder with the weight limit of your bot. You obviously cannot use a 50-pound cylinder on a 60-pound robot. Once you determine the amount of CO_2 or HPA your bot can carry, pick a cylinder size based on the amount of force you want to generate. Do the calculations in the pneumatics chapter to estimate how many shots you will get out of the weapon. If the number is too small, change the cylinder size. If the number is really big, either double the volume of the cylinder for the return stroke or move the cylinder size up some to gain more force.

Q. Is this valve, regulator, or cylinder any good?

A. I don't know. I am definitely not a pneumatics guru. However, there are several Web sites with multitudes of information based specifically on using pneumatics in battling robots. Most teams like to keep their choice of valves secret, but many do not mind sharing the specs of the regulator and cylinder they like.

Q. How much force will I get?

A. Force equals pressure times area. Area equals half the bore squared times π.

Force = Pressure x Area

Area = $(\text{bore} / 2)^2 \times 3.14$

$A = (3/2)^2 \times 3.14$

$A = 2.25 \times 3.14$

A = 7.07 square inches

$F = P \times A$

$F = 150 \times 7.07$

F = 1060.5 pounds

If you have a 150-psi system and a 3-inch bore, 6-inch stroke cylinder, the area is 7.07 square inches and the force works out to be about 1,060 pounds. Do not count on getting that amount of force every time. Depending on how you are using the cylinder, you will never get that amount of force. Remember that any kind of lifting device that is attached to the cylinder forms a lever. Levers exchange perceived weight for distance. If you have your cylinder pushing up on a set of lifter arms that are 12 inches long, you must divide the force by 12 inches to find the actual amount of force seen at the ends of the arms. In this case, you would be lifting with about 88 pounds of force. If you change the geometry of the lifter arms, you start getting a lot of that force returned to you. A 6-inch lifter arm will produce about twice as much force.

Q. How fast will it work?

A. Gas expands really quickly. That is why pneumatic systems are so dangerous. If you are experiencing slow actuation times, 1 second or more, then there is something wrong with your design. It may be the parts you are using. You want to have the highest flow rate valves you can get your hands on. Large ID tubing is a must for good flow rates. Most regulators, if not all, are not designed for high flow rates. Using a buffer tank directly after the regulator is a good idea.

Q. What do I need for Valves?

A. Use only professionally built, industry rated parts. Homemade pneumatic parts are not for the new builder. Some people have tried to use lawn sprinkler valves. Those are not designed for what we are doing here. Besides being plain dangerous, they are against the rules of most if not all competitions.

Q. How many hits will I get from a tank of CO_2 or HPA?

A. I explain this in greater detail in the pneumatics section of the book, but here goes, using 150 psi CO_2, a 20-ounce paintball tank, and a 3-inch bore, 6-inch stroke cylinder as the example. The size of the tank in ounces is the weight of the CO_2 you have when the tank has a full charge; 20 ounces is about 1.25 pounds, and 1 pound of CO_2 expands to about 8 cubic feet at 14.7 psi or normal air pressure. So, 1.25 pounds of CO_2 expands to about 10 cubic feet at 14.7 psi. You are not using the gas at 14.7 psi but at 150 psi. Convert the known volume (10 cu ft) and pressure (14.7 psi) to the volume at 150 psi by using Boyle's law. Pressure^1 times Volume^1 equals Pressure^2 times Volume^2. Solve the equation and get almost 1 cubic foot for the second volume. Convert 1 cubic foot to get 1,728 cubic inches. Calculate the volume of the air cylinder by multiplying the area times the stroke length. That volume equals about 42.42 cubic inches. That is how much it takes to push the cylinder out. Double it and get 84.84 cubic inches to estimate the amount of gas required to fully extend and fully retract the cylinder piston. Divide the total volume of gas you have (1,728 cu in) by the amount you need per shot (84.84 cu in) and the answer is about 20 shots.

Q. How do I build a system that can throw an opponent across the arena?

A. You have to overcome gravity. Gravity pulls down on your opponent with a specific amount of force. Depending on how far your lifting arm will be lifting and how high you want to toss the other bot, you have to use a proportional amount of force. For instance, if you want to toss the bot 6 feet up and you are lifting, for 1 foot you must put 6 times the weight of the bot in as force.

Radio Control Questions

Q. PCM radios are really expensive. Why can't I use a regular FM radio?

A. PCM radios are much safer to use for robot control than standard FM radio gear because they filter out interference. Interference is caused by the motors in your robot and any electrical equipment inside or near the fighting arena. The real question you need to ask yourself is "Can I afford the personal injury to myself or others that will result if my bot runs out of control when it is not supposed to?" You can do everything known to man to reduce interference, but the nature of an out-of-control robot is unpredictable. That means you don't know when it's going to take off and chase someone. There is another choice in the matter. It is called IPD technology and is outlined in Chapter 2. It is operated on standard FM radio gear, yet the receiver is specialized so that it can handle interference in a manner similar to a PCM radio. People have tried it and, as of this writing, the jury is still out on whether or not it works as well as a PCM setup. Some competitions do not allow IPD radios.

Q. How do I make my robot turn? How do I steer my robot?

A. The answer to this question depends, yet again, on the design of your robot. It is covered in more detail in Chapter 2. In short, many competitors use "tank steering" to drive their bots, where each side of the robot is independently controlled. To turn right, the left side of the bot is commanded to go forward, while the right side is commanded to either go in reverse, to stand still, or go forward at a slower rate than the left side. Mixers combine the channels of the RC setup so that control is intuitive on a single joystick.

Q. What do I need for controls?

A. The main control components of your robot are the remote control unit, the mixer (optional), the speed controllers, the speed controller interface (optional depending on the controller), the motors, and the batteries. Which type of components you need depends on many things. Chapters 1, 2, 4, 5, and 6 detail what has been used in controlling bots. Do some Internet research on your own and make your decisions wisely based on the information you find there and in this book.

Q. Do I need to use four RC channels when using four motors in the drive system?

A. The standard drivetrain setup of a bot using four motors only requires two RC channels to control it: two motors on each side of the robot running at the same speed and direction. One channel can be used to control one side of the bot. If you need to run all four motors at different speeds and directions, then you will need four channels and four speed controllers to control it.

Q. Where do you mount the antenna on a spinning-shell robot?

A. To get the best range from your radio system with this type of bot requires some creative design work. One possible way is to have the antenna mounted to some part of the shell that doesn't spin. For example, you can have the shell spin around a central, nonspinning, hollow post. The antenna can be mounted inside or on top of that hollow post. Also, depending on the shell itself, you may be able to mount the antenna inside the shell. If the shell is completely metallic, you are likely to suffer from very poor or reduced radio control range. If you are going to build one of these robots, test it thoroughly with different antenna positions.

Weapon/Armor Questions

Q. Is (insert material here) good armor for a bot?

A. Many materials may be considered good armor. Some materials are too heavy to be used. Some materials do not have the structural properties that are required to be good armor. Unless you are a materials engineer with time to test and money to purchase exotic materials, I suggest you stick with what has been traditionally used on past robot champions. Chapter 11 and Appendix C cover several of the popular materials and their pros and cons. If you don't see the material listed that you had in mind, chances are that it's not really a good candidate. If you don't believe me, prove me wrong by using it in the next competition. Let me know how it works out.

Q. Is (insert gizmo here) a good weapon for my bot?

A. I don't know. There are too many unknown factors to answer that question. Chapter 12 covers many of the popular types of weapons. If you have a new idea for a weapon, build it and see how it fairs against targets and opponents. Take a look at past competitors to see how they did with a particular weapon design. See if you can improve on their design or incorporate their improvements in yours.

Q. Can I use a big magnet?

A. Why? I'll take a gamble and say that I don't believe anyone can get enough current out of the number of batteries they can fit in the weight class for the length of time it would take to power a big enough electromagnet to pick up and carry an opponent to an arena hazard. Besides, magnets require a big flat space to get a solid hold. They also require the opponent to be made from steel. I've seen 325-pound robots that were made of mostly polycarbonate and aluminum. What good would a magnet do on that? Some bots have tried to use magnets to gain attraction but they still need work.

Construction Questions

Q. How much pushing power do I need for my bot?

A. Pushing power is relative to traction. If you don't have traction, you aren't going to push anything around because your wheels will spin. Most people don't design according to pushing power. Most people design according to how much weight the robot is able to carry. They also include the ability to get under an opponent. If you can get under your opponent, you will increase the amount of traction and therefore increase the actual pushing power you have if the weight doesn't overpower your motors.

Q. How fast should my bot be able to go?

A. The easy answer is that your bot should be able to go as fast as you can comfortably control it. If you are looking for a hard number to shoot for while designing, then I'd say between 10 and 20 feet per second. That works out to be between about 7 and 14 miles per hour. If your bot has a weapon that needs aim to hit effectively, you want to go slower than a bot who just rams into its opponent. Remember that you trade the power necessary to push people around for speed.

Q. Which is better—chains, gears, or belts?

A. Each has its own pros and cons. Chapters 6 and 7 explain chains, gears, and belts.

Q. How big do my wires have to be?

A. Required wire size can be determined accurately by knowing the amount of current that will flow through them. You should know the maximum amount of current that your motors draw and use wires that can handle it. There is a chart of wire size and current carrying capability in Appendix C of this book. If you cannot get the size of wire that is specified, it is possible to share the current over two wires that are half (or bigger) the required size. This is not recommended because of reliability reasons. You can guess the amount of current a motor was designed to use by the size of the wires it uses from the factory. However, this might not be an accurate rating since most factories aren't going to design a product that runs a motor at or near the stall ratings, which it might easily see in a fighting robot.

Q. How do I connect my wires together?

A. There are literally thousands of different types of wire connectors. When choosing a connector keep two things in mind: current rating and connection strength. The connector must be rated to handle the amount of current the wire carries. Connection strength is important because you don't want the two connectors to separate when the robot takes a big hit from an opponent. If you can't get connectors that take some effort to separate, you can possibly strengthen the connection using hot glue, tape, wire ties, etc.

If the wires are connecting to a terminal strip (strip with screws in it), then you'll need some form of spade or ring connector. I recommend the ring connectors, since they will not fall off if the screw vibrates loose.

In emergency repair situations, I've used wire-nuts to join two or more wires together. This is a bad idea if you can avoid it and should only be done as a last resort. If you have to use them, wrap them and the wires with electrical tape. A better idea is to use two ring terminals and bolt them together. Cover that with electrical tape to avoid shorts.

Q. What kind of metal is good for a lightweight bot?

A. The answer depends on which part of the bot you are asking about. If it is the frame or supporting body, aluminum has been used with great success. If it's the active part of the weapon, then you may need to use something harder. That really depends on the design of your robot and weapons. You wouldn't want a spinning blade weapon made from aluminum because it would bend way too easily. I've had a 3/8-inch thick, spinning steel blade weapon bend before the end of every match the robot entered. However, some supporting parts of the weapon may very well be able to handle the stresses incurred, even if they are made from aluminum.

Q. It's really hard to figure out how big some bots are because I've never actually seen a competition live, so what is the average size of lightweight bots (length, width, height)?

A. As far as I know, there is no average size. Competitions that have the lightweight class usually require the robot to weigh at most 60 pounds. I've seen 60 pounds of robot crammed into a 13-inch diameter × 6-inch height. I've also seen 60 pounds of robot spread out over a 4- or 5-square foot area. The best thing to do is get your hands on some parts. Weigh and measure them. Then design your bot according to those sizes and weights. This will tell you about how much your bot will weigh when you are done. Many parts are listed in catalogs along with their specifications, so that you don't need to buy them in order to design them.

Miscellaneous Questions

Q. What's the cheapest speed control solution for my robot?

A. The answer depends on the motors and drivetrain that you will use. Check the speed controller section of this book and do some research on your own as well. Get a speed controller that will handle the amount of current asked for by the motors. If you don't feel that you need actual speed control, you can use a relay or solenoid type H-bridge that simply turns your motors on and off in either direction. However, this setup can cost as much or more than a true speed controller and offers much less control over your robot.

Q. Can I use automotive relays, contactors, or solenoids to control my motors?

A. It will work, but this setup will only turn the motors on and off. They either run at full speed or are off. You will need some method of converting the receiver signal into a signal that will turn the relays, contractors, or solenoids on and off. These will cost between \$30 and \$40 from a reputable robot parts Web site. The problem with the automotive relays, contactors, or solenoids that cost \$8 is that they aren't for continuous duty and some don't hold their contact when your bot gets hit hard by an opponent. Other contactors do the job but cost more. I've bought some that handle 500 inrush and 200 continuous amps but they cost about \$40 each with a discount. You'd need eight so that's \$320 plus \$60 for the converters. That's \$380+ for on/off control. You can pay about \$300 for controllers that handle EV warrior motors, cordless drill motors, and many wheelchair motors *and* give you actual speed controls. So, if you are using any of the three types of motors I mentioned (someone asking this question is most likely using these), there's no reason to build a controller based around relays, contactors, or solenoids.

If you built an actual speed controller with MOSFET technology, it would probably burn up many times before you got it to work right with the motors you want to use. Many people have tried. Even the experts that I've seen try it say that it's not worth building yourself. You can either take their advice and save some money or build it and see.

Q. When and where is the next competition?

A. Quite a few small competitions are springing up. The best way to answer this question is to do some research on the Internet. All of the competitions that I know of have a Web site. Many of the Web sites link to other competition sites. Start looking at whichever sites are mentioned by the TV shows you like.

Q. Is (insert name) available for use as my robot's name?

A. There is no official listing of names besides the competitor lists at a competition's Web site. Check there for the name in which you are interested. There are a few robot databases on individual builder's Web sites. They are easy to find once you start looking, and I'd rather not list specific Web sites due to the volatile nature of the Internet.

Q. Wanna hear about the awesome idea I have for a bot?

A. Although I live and breath robots, I cannot personally critique everyone's design plans. There just isn't enough time in the day for that and my own robot building exploits.

Q. What kind of (component) does (robot) have?

A. This question comes up many times. It is a very bad question to ask for someone wanting to learn about building robots. It is not a bad thing to want to know the answer, but asking for an answer when you can easily find the answer by doing some research is just lazy. The best thing to do is find the Web site of the robot you are talking about and read everything the builder has to say. If the robot doesn't have a Web site or the site doesn't mention what you want to know, find out who built it and email him asking the question.

Q. How do I get a sponsor?

A. This is covered in some detail in Chapter 17.

Mechanical Questions

Q. What is the best type of steel to use for axles?

A. I've never used anything other than the common steel found at local machine shops. Doing so requires that you do not skimp on the amount of material. See the section about axles. I know that some people have used 304 stainless steel, 01 drill rod, and keyed go-cart axle. Remember that tool steel is hard and brittle and not very fatigue resistant.

Q. How does one prevent an axle from moving back and forth in its bearings?

A. There are several ways to accomplish this. You can fabricate your shaft so that it has a shoulder that rides up against two bearings. Some people machine grooves for snap rings in the ends of the axles. The easiest way to control the movement is to use a couple of shaft collars.

Q. What size of threaded hole is safe to place in a certain thickness of material?

A. The rule of thumb is 1 1/2 times the diameter of the hole from hole center to edge. So, if you have a 1/2-inch thick piece of material that needs a tapped hole in its edge, the hole should be no larger than 0.1667 inches or about a 10-32 bolt. If this seems small, you can probably get away with a 12-24. Be sure to use multiple bolts so that they can share the load.

Q. I'm trying to use a chain drive system but the chain keeps coming off the sprockets. What can I do?

A. Sprockets must be aligned correctly. To do this, place a straightedge flat against both sprockets. If there is any light between the straightedge and either sprocket, they are not aligned correctly. Move one or both sprockets to where there is no light showing between them and the straightedge.

Also, chains and sprockets should not to be mounted in a vertical manner. That is, one sprocket should always be to the right or left of the other and never straight above, because chains tend to stretch when in use and could start hanging off the teeth on the bottom sprocket. Using an idler sprocket may help if you do put one sprocket directly above another.

Make sure to have some kind of adjustment in your drive train so that when the chain gets loose you can either move the motor or an idler sprocket to take up the slack. Remember that even if you use an idler sprocket you can still have problems. The chain is only tight on the side that is being pulled. So, if there is slack in your system, the chain can droop on the side being pushed if the idler is on the driven side. This is apparent in robots, because we typically have the motors reverse directions. The tight and slack sides are constantly changing hands. A two-idler system is best in this case.

GLOSSARY

The following definitions pertain to how each word is used in this book and the fighting robot community and are not necessarily the *Webster* approved definition.

Abrasive wheels
Hard abrasive (scratchy) wheels used for grinding.

After-market
Products made by companies other than the original equipment manufacturer.

AC
Alternating current. Type of electric current that reverses its direction of flow. Most commonly known as house current.

Acme
Screw thread form.

Alloy
Two or more metals combined to form a new metal.

AM
Amplitude modulation. A method of sending radio signals over the air to a receiving device. Also, a method of varying the amplitude of a frequency to encode a message on a carrier wave.

Amp
A measure of electric current.

Amp-hour
A measure of battery life indicated by how many amps it could supply over a certain period of time.

Angle iron
An iron or steel structural member that is L-shaped.

Annealing
The heating and cooling of a metal to make it easier to work with.

Anode
Positive terminal of diode.

Anodize
Using an acid bath to make aluminum form a very hard aluminum oxide layer.

Arbor
A shaft for holding cutting tools.

Arbor press
A hand-operated machine tool used to apply high pressure for pressing together or removing parts.

Atom
Smallest particle having the characteristics of an element; protons, electrons, neutrons.

Autonomous robot
A self-guiding robot. A robot that makes it's own decisions and interacts with it's environment.

AWG
American Wire Gauge. A standard of measurement.

Axle
A shaft that is driven usually to turn a wheel.

Babbitt
A metal alloy used for bearing inserts.

Backlash
The looseness or play between the faces of meshing gears or threads.

Ball bearing
A style of bearing that uses balls to eliminate friction.

Bandsaw
A saw whose blade is a continuous, thin, steel band having teeth on one edge and passing over two large pulley wheels.

Bang-bang
A method of controlling electric motors by turning it on and off at full speed.

Bar stock
Metal bars of various lengths and shapes from which parts are fabricated.

Bastard
A standard coarse-cut file.

Bearing
A part that uses balls in a track to ease motion.

Bench grinder
A small grinding machine for shaping and sharpening the cutting edges of tools.

Bench vise
A strong device mounted to a stand or a bench; meant to be used to hold parts.

Bevel
Surface slanted to another surface.

Biped
Having two legs or feet for walking.

Bit
A hardened steel tool used to cut materials.

Blind hole
A hole that does not pass through the material in which it is drilled.

Body
The main section of the robot. Also the small section of a rivet left in the two parts being fastened.

Bolt circle
Center line circle for locating holes around a common center point.

Bore
To enlarge and finish the surface of a cylindrical hole. Also refers to the diameter of the round hole in a part such as a gear, sprocket, pulley, or cylinder.

Boring tool
A cutting tool in which the tool bit, the boring bar, and, in some cases, the tool holder are incorporated in one solid piece.

Boss
Raised surface of a circular outline.

Boyle's law
A rule stating that the pressure times the volume of a gas that expands equals the new pressure times the new volume.

Brass
Alloy of copper and zinc or copper with zinc and lead.

Braze
To join two pieces of metal by using a hard solder like brass or zinc.

Brittle
Opposite of ductile. Does not bend much before breaking (like glass).

Broach
A tool used to cut slots in material.

Bronze
Alloy of tin and copper.

Brushless motor
A motor that does not cause radio frequency interference due to the making and breaking of brush contact.

Buff
To polish to a smooth finish.

Buffer tank
A device used to contain expanding gasses used in pneumatic systems.

Burnishing
To make smooth or glossy by rubbing.

Burr
Rough edge caused by the fabrication process.

Burst disk
A safety device mounted on the neck of a tank of supply gas, CO_2 or otherwise, for the purpose of expelling gas if a safe level of pressure has been exceeded.

Bushing
Hollow, round sleeve used as a bearing or drill guide.

CAD
Computer Aided Design. Sometimes also Cardboard Aided Design or Chalk Aided Design.

Caliper
A device used to measure inside or outside dimensions.

Cam
Machine part that changes rotary motion to linear motion.

Camlock bearings
Bearings that squeeze the shaft in order to lock into place.

Capacitor
An electrical device used to filter noise from the robot's power lines.

Carbide tool bits
Cutting tools to which carbide tips have been attached to provide cutting action on harder materials.

Carbon steel
A broad term applied to tool steel other than high-speed or alloy steel.

Carburetor
A device used for vaporizing a fuel to produce a combustible gas.

Case hardening
A heat-treating process that makes the surface layer or case of steel harder than the core.

Caster wheel
A free rolling wheel used to balance two-wheeled bot whose center of gravity is not centered over the drive wheels.

Cathode
Negative terminal of diode.

Cell
A single battery that usually has a 1.2 voltage rating.

Center of gravity
The point in a robot where the weight is balanced.

Center punch
A pointed tool made of hardened steel, used to mark the workpiece with a dot or dent.

Chain gearing or drive
Power transmission by means of chain running around sprockets.

Chamfer
Edge that has a bevel.

Channel
A frequency of transmission used by remote control systems. Also, individual streams of information sent by the transmitter and interpreted by the receiver.

Chassis
The frame of the robot.

Chisel
Any one of a variety of small, wedge-shaped cutting tools.

Chuck
A device on a machine tool to hold the workpiece or a cutting tool.

Circuit
A closed path providing continuous passage of fluid or electricity.

Circular pitch
The distance measured on the pitch circle from a point on a gear tooth to the same point on the next gear tooth.

Circumference
The distance around the outside edge of a circular part.

Clamper
A type of robot that grabs and holds on to its opponent.

Clamp-on amp-meter
A meter used for measuring amperage without being inserted into the circuit.

Clearance
The distance or angle by which one object or surface clears another.

Cold-rolled steel
Steel that has been rolled to accurate size and smooth finish when made.

CNC
Computer Numerical Control. Used with automatic milling and lathe machines.

Common
As referred to in a switch, the terminal that will connect to either the normally open or the normally closed terminals.

Compressed gas
A method of powering certain weapon systems.

Computer radio
A remote control unit that uses a computer to adjust and remember settings.

Conductor
A material that allows electrical energy to pass.

Constant current
A method of battery charging that keeps the amount of current flowing constant.

Constant voltage
A method of battery charging that keeps the amount of voltage constant while the current varies.

Contactor
A large switch that is actuated by electromagnetic energy.

Contact patch
The area of contact between a tire and the driving surface.

Counterbore
Enlarged end of hole to some depth.

Countersink
Cone-shaped space at the end of a hole.

Coupling
Mechanical attachment of a load to the motor or a shaft to a shaft.

CO_2
Carbon dioxide. Used in some pneumatic systems in combat robots.

Crimper
A device used to mechanically connect a terminal end to a wire.

Crystal
A device used to regulate the frequency of a radio transmitter and receiver.

Current
A moving stream of water, air, electricity etc.

Current limiting
A protective feature found in some speed controllers that keeps the amount of flowing current under a threshold set by the current handling abilities of the controller itself.

Cutting fluid
A liquid used to cool and lubricate the work and cutting tool.

Cutting speed
The surface speed of the workpiece.

Cylinder
An enclosed piston with a hole on each end. The piston is moved by pumping fluid or gas into an end.

DC
Direct current. Electric current that always flows in one direction. Most common in batteries.

DC-DC converter
An electric device that converts one voltage to a constant lower voltage.

Deburr
To remove sharp edges.

Density
The measure of closeness of small parts in a large part. Usually used when describing metals.

Diameter
The length of a line drawn from a point on a circle through its center to the opposite point on the circle.

Die
Cuts external threads of a screw and can also be a punching tool.

Diode
Electrical device that allows current to flow in only one direction.

Discharge current
The amount of current drained from a battery.

Double acting
An air cylinder that must be pressurized in order to be extended or retracted.

Double-stage reduction
A gear train in which the motor's speed and torque are reduced through two stages of gears to get a higher torque and slower speed than would be accomplished with one pinion and drive gear.

Dowel
Cylindrical-shaped pin used for fastening parts together.

Drill
A pointed, round tool that is rotated to cut round holes in material.

Drill chuck
Used to grip drills and attach them to a rotating spindle.

Drill press
Machine for drilling holes in metal, wood, or other materials.

Drill rod
A high-carbon steel rod with a smooth finish accurately ground to size.

Driven gear
The gear in a gear reduction system that is driven by the pinion.

Drive train
The mechanical parts of a robot that are directly responsible for movement.

Dry cell
A battery that contains dry chemicals instead of liquid or gel.

Ductility
The property of a metal that permits it to be bent without cracking or breaking.

Duty cycle
The planned amount of time spent working or running.

ECRA
East Coast Robot Alliance. A group formed by East Coast builders to arrange for better transportation prices and to arrange East Coast robot competitions.

Elasticity
The measure of a materials' ability to revert back to its original form after being deformed.

Elastic deformation
Reversible deformation; like a spring.

Electric actuator
A device that extends and retracts using electrical energy rather than pneumatic or hydraulic energy.

Electrode
A conductor of electricity found inside a battery.

Electromechanical
Using electricity to affect something mechanically.

Electrolyte
The acid mix inside a battery.

Electrolytic capacitor
A capacitor that has specific positive and negative leads.

Electron
The orbital part of an atom. Has a negative charge.

Elevon mixer
An electronic device that mixes radio control signals so that it's easier to drive a robot with one joystick.

Emergency cutoff loop
An electrical device used to break a circuit in a hurry.

Epoxy
A high-strength glue.

Extruded
Metal which has been shaped by forcing through a die.

Extrusion
A shaped part resulting from forcing a material through a die opening.

Face
To cut a flat surface. Also, the width of the mating part of gears.

Failsafe
A device or method of programming that sets the control of a robot back to a safe position if radio contact is lost.

FCC
Federal Communications Commission. The commission that sets the rules that govern the use of radio waves.

Ferrite core
A ring or rod made of a type of iron.

Fillet
Rounded corner between two surfaces.

Fit
Tightness or looseness between parts.

Fixture
Device used for holding a workpiece. *Also see* Jig.

Flange
Extension from a surface.

Flipper
A type of robot that flips its opponents over in order to win.

Flux paste
A pasty substance that helps solder to stick to a part when applied along with heat.

FM
Frequency modulation. A method of sending radio signals across the air to a receiving device. Also, a method of varying the frequency of a radio wave to encode a message on a carrier wave.

Force fit
A fit between parts where one part is forced or pressed into another.

Frequency
The rate at which something happens; specifically, the rate of radio waves sent by the transmitter.

Fulcrum
The point or support on which a lever turns.

Gait pattern
The pattern formed by mapping the points of contact between walking feet or legs and the contacted surface.

Gauge
Device for determining if a dimension on an object is within specified limits. Also, a standard of measurement.

Gear ratio
Ratio of motor rpm to wheel or axle rpm.

Gear reduction
Using gears, sprockets or pulleys to change the speed and strength of a motor.

Gel cell
A popular battery that does not leak when upside down or cut.

Ground
A path of electrical current to the Earth or a conductor of equivalent effect.

Ground isolation
The practice of keeping two interacting systems with different voltages electrically separate.

Gusset
Extra material placed in strategic places on the frame to add strength and rigidity.

Gyro
A device used to alter signals between the receiver and the speed controller so that the robot will be more apt to run in a straight line.

Hacksaw
A metal blade with small, close teeth on one edge.

Hardening
A process for steel that increases the hardness and tensile strength and reduces its ductility by applying heat.

Hardness
The measure of a material's ability to resist denting.

Heat
Form of energy.

Heat treatment
Heating and cooling a metal to change properties or characteristics.

Helical gear
A gear with teeth cut at an angle other than a right angle across the face of the gear.

Hex
A term used for anything shaped like a hexagon.

High-speed steel
An alloy steel commonly used to make cutting tools.

Hole saw
A cutting tool used to cut a circular groove into solid material.

Honing
Finishing surfaces to a high degree of accuracy and smoothness.

Horsepower
The measure of power output of a motor.

Hot-rolled steel
Steel that is rolled to finished size while hot.

HPA
High pressure air.

Hub diameter
The diameter of the hub of a gear, sprocket, or pulley.

Hub projection
The distance that the hub material extends past the tooth material on a gear, sprocket, or pulley.

Hydraulic
Powered using some sort of fluid.

ID
Inside diameter.

Idler
A gear placed between two other gears to transfer motion from one gear to the other gear without changing their speed or ratio. Also used with sprockets and pulleys to take the slack out of the chain or belt.

Inch-pounds (in-lb)
The unit measure of torque. Also known as pound-inches (lb-in).

Innovation First
A company that builds and sells robot radio and speed controllers.

Insulator
A material that does not allow electrical energy to pass.

Invertable
Having the ability to work as designed while inverted; running while upside down.

Inverter
An electronic device that changes direct current electricity into alternating current electricity.

IPD
A style of encoding that sort of mixes PPM with PCM remote control transmission.

Jig
A guide for a cutting tool. *Also see* Fixture.

Kerf
The width of cut made by a saw.

Key
Small metal object designed to fit key seats in a shaft and hub of a gear or pulley to provide a positive drive between them.

Key seat or keyway
Groove or slot in a shaft or hub into which a key is placed to give positive contact during rotating force.

Kill switch
A safety device installed in robots that can cut all power to the drive and weapons systems.

KISS
Keep It Simple Stupid. An engineering motto meant to remind designers not to complicate things any more than necessary.

Knurl
Series of cuts to roughen a round surface so that it can be more easily turned by hand.

Length through bore
LTB. The length of the cylinder formed by a hole in a part.

Lever
A device used to put leverage, lifting pressure or otherwise, on another object.

Lifting arm
A weapon used by robots to pick up and/or lift an opponent.

Linear actuator
A device that translates rotary movement to linear movement.

Load
That which is carried. The amount that can be carried or contained.

Lockout
A condition that happens to PCM receivers either when radio contact is lost or when significant interference is received.

Machinability
The degree of difficulty with which a metal may be machined.

Magic smoke
The smoke that comes out of any electronic device when it is destroyed, either through mechanical or electrical means.

Magnesium
A lightweight, ductile metal similar to but lighter than aluminum.

Main power switch
The switch in the robot that kills power to all drive and weapon systems.

Malleable
Capable of being extended or shaped by hammering or rolling.

Mandrel
The part of a rivet that looks like a nail.

Maximum bore
The largest bore allowed, in a part, which is allowed by the amount of material present.

Mic or mike
Machinist's slang for micrometer or to measure with a micrometer.

Mild steel
Low-carbon steel.

Mill
To machine a part on a milling machine.

Miter
45-degree bevel.

Modulus of elasticity
Ratio of stress to strain in elastic region; stiffness of spring.

Moment of inertia
Mathematical construct that allows the geometrical shape to indicate strength.

Morse taper
A self-holding standard taper.

Motor controller
A device that allows an output from an RC receiver to drive an electric motor. Can also be known as a speed controller, yet a motor controller doesn't always control the speed of a motor.

Mounting distance
The distance measurement between the centers of two meshed gears.

Neutron
Part of the center of an atom. Has no electrical charge.

NiCd
Nickel-Cadmium. A type of dry cell battery.

NiMh
Nickel-Metal Hydride. A type of dry cell battery.

Nominal
Usual.

Nonferrous
Metal containing no iron (brass and aluminum).

Normally closed (NC)
The terminal of a switch that is connected to the common terminal until the switch is actuated.

Normally open (NO)
The terminal of a switch that is not connected to the common terminal unless the switch is actuated.

Nucleus
The center of an atom made of protons and neutrons

Ohm's law
Voltage (V) equals Current (I) times Resistance (R) or $V = IR$.

Oil-lite bearing
A bearing made of bronze alloy that contains oil.

Outside diameter
The length of a line drawn from a point on the outside circle of a round part through the center to the opposite point on the outside circle.

Overhung axle
An axle that is supported in two places but the driven part is not between the two supports.

Pack
A group of individual battery cells connected and bound together to form a specific voltage and current rating.

Parallel
A method of wiring that connects the positive terminals of two motors or batteries to each other and the negative terminals to each other.

PCM
Pulse Code Modulation. A method of encoding radio signals so that interference problems are kept to a minimum.

Peak current
The highest amount of current expected.

Pillow block
A bearing mounting style that positions the bearing perpendicular to the mounting surface.

Pilot hole
A small hole drilled in a part so that it's easier to drill a large hole.

Pinion
The gear, in a gear reduction, that is spun by the motor.

Piston
The plunger-like device in an air cylinder that extends or retracts when the system is pressurized.

Pitch
The length between two teeth on a gear, sprocket, or pulley.

Pitch circle
A theoretical cylinder that passes through the threads in such a position so that the widths of the thread ridges and grooves are equal in dimension.

Pitch diameter
The length of a line drawn from a point on the pitch circle of a round part through the center to the opposite point on the pitch circle.

Pivot point
See Fulcrum.

Plain bearing
A style of bearing made from a bronze alloy but does not contain oil like the Oil-Lite style bearing.

Plastic deformation
Non-recoverable deflections (bending).

Plate
To electrochemically coat a metal object with another metal.

PMDC motor
Permanent Magnet Direct Current Motor. A type of motor that has permanent magnets mounted on the inside and runs on DC electricity.

Pneumatic
Powered using air or other gas.

Polarized
Having different qualities in different directions. In capacitors, having specific positive or negative properties.

Polycarbonate
Strong, bullet-resistant, usually clear material popularly used as armor. Also known as "bullet proof glass."

Power
Time rate of doing work.

PPM
The cheapest style of FM modulation in commercial hobby remote control units.

Proton
Part of the center of an atom. Has a positive electrical charge.

PSI
Pounds per square inch. A measure of pressure.

Pulse Width Modulation (PWM)
The practice of changing the width of a train of pulses to represent information.

Punch
To pierce thin material by pressing a tool through it. The tool is usually called a punch as well.

Race
The part of a bearing that the ball or roller is contained within.

Rack
An array of gear teeth on a straight bar.

Radio Frequency Interference
Noise that is introduced into the waves of a radio transmission.

Radius
The length of a line drawn from the center of a round part to a point on the outside edge of the part.

Rammer
A type of robot whose primary weapon is to run into its opponent and do damage.

RC
Short for Remote Control or Radio Control. A remote control transmitter and receiver set.

Ream
To make a hole the exact size by finishing with a rotating, fluted cutting tool (not a drill bit).

Reduction ratio
The value obtained when dividing the driven gear's number of teeth by the pinion gear's number of teeth.

Regulator
Controls the output pressure of a pneumatic system.

Relay
A switch that is actuated by electromagnetic energy. Large ones are sometimes called contactors.

Remote control receiver
A device used to receive radio signals from the driver of the robot.

Remote control unit
The equipment used to control a robot from a distance.

Resistor
An electronic device used to limit the amount of current flowing in a circuit.

Resistor spark plug
A special spark plug that reduces radio frequency interference.

RFI
Radio Frequency Interference. Caused by electric motors, lights, and other devices.

Roll pin
Hollow circular piece of steel.

Roller bearing
A style of bearing that uses rollers to eliminate friction.

Rotor
Part of the motor that spins.

RPM
Revolutions per minute. A measurement of rotational speed.

Safety factor
Ratio of material's ultimate stress to the highest calculated stress.

Scale
A shop term for steel rulers.

Schematic
A drawing of an clectrical circuit.

Self-righter
A type of robot that has the ability to right itself if it has been turned upside down.

Series
The method of wiring motors or batteries in which the positive terminal of one battery or motor is connected to the negative terminal of the other battery or motor.

Series wound motor
A motor in which the magnets are replaced by more windings.

Servo motor
A geared motor that includes a closed-loop control system that allows precision positioning of the control horn, or arm. Also, a DC motor with high torque and low rpm.

Set screw
A small, threaded part used to secure bearings, sprockets, gears, pulleys, and other things to shaft materials.

Shear
To cut material between two blades or other materials.

Shear modulus
Materials resistance to shear forces.

Shim
Thin material used between two surfaces to adjust the distance between them.

Shock mounting
A method of mounting components so that they do not feel significant vibration due to impacts.

Short circuit
A low-resistance electrical connection that is usually an accident.

Shrink wrap
A rubber-like material that shrinks when heated. Used to insulate bare connections.

Shufflebot
A robot that uses elongated feet rotating continuously around a cam-type drive device to walk or shuffle across the driving surface.

Shunt
A calibrated resistor used to measure voltage drop to calculate the amount of current used by a system.

Single acting
An air cylinder requires pressure to extend but requires forces other than pressure in order to retract.

Single pole double throw
A switch designation showing that the switch has one common connection, a normally open connection and a normally closed connection.

Single-stage reduction
A gear train in which the motor's speed and torque are reduced through one stage of gears to get a higher torque and slower speed than would be accomplished by mounting a wheel directly to the motor shaft.

Skid steering
A method of driving a robot in which one or more wheels slide across the floor instead of turning.

Skirt
A form of armor mounted so that it drags on the floor to help ensure the opponent does not get under the robot.

SLA
Sealed Lead Acid. A type of battery in which the electrolyte cannot escape. Also known as a gel cell.

Solenoid valve
An electric valve used to control hydraulic or pneumatic systems.

Solid-state
Being completely electronic in nature.

SORC
Society of Robotic Combat.

Specific gravity
The ratio of a material's density to the density of water.

Specific volume
Volume per unit mass.

Speed controller
A device used to control the RPM of a spinning motor or the velocity and direction of a robot.

Spindle
A rotating device used in machine tools such as lathes, milling machines, and drill presses.

Spindle speed
The rpm at which a machine turns its spindle. *See* cutting speed.

Spinner
A type of robot that either spins or has a spinning weapon.

Spline
Long keyway (or set of keyways) down a drive or motor shaft.

Sprocket
A gearlike device used for chain-driven systems.

Spur gear
A gear having teeth parallel to the axis of its shaft.

Stall current
The amount of current drawn by a motor when the shaft is held in one position and full voltage is applied.

Stall torque
The amount of torque created by a motor when the shaft is held in place and full voltage is applied.

Stepper motor
A motor that has several windings that have to be energized in a certain order to rotate the shaft.

Sticky back
A plastic device used in conjunction with wire ties to hold bundled wires in one place inside the bot.

Strain
Physical changes in a material resulting from stress.

Stress
Force over an area.

Supported axle
An axle that has support on both ends and the driven part between the supports.

Tachometer
A device used to measure the number of revolutions completed by a spinning object.

Tank steering
A method of steering a robot in which each drive wheel is controlled separately.

Tap
To cut threads in a hole.

Tap and die set
A set of tools made to cut threads in material.

Taper
A uniform increase or decrease in the size or diameter of a workpiece or tool.

Temper
To reduce, by heating, the brittleness in metals that have been hardened.

Tensile strength
The measure of a part's ability not to break under pressure.

Terminal
An electrical connection or connector.

Thread
A helical projection of uniform section on the internal or external surface of a cylinder or cone.

Timing belt
A toothed belt used because of its ability to resist slipping.

Tin
To lightly cover something with solder.

Tolerance
Specified allowable variation from a given dimension.

Tool steel
High-carbon steel that can be heat-treated to a hardness required for metal cutting tools.

Torque
A measure of force that a motor can apply.

Torque constant (Kt)
How many amps it will take for a motor to produce a specified amount of torque.

Toughness
How much energy it takes to go from yield to fracture.

Trickle charge
Charging a battery at a very slow rate.

UHMW
Ultra High Molecular Weight. A plastic-like material.

Ultimate strength
Stress at point of fracture.

Ultimate stress
Highest stress a material can withstand before failure occurs.

Variable resistor
A device used to change the resistance to electrical current flow.

Vantec
Name brand of a popular speed controller.

Volt
A measure of electric potential.

Voltage drop
When the voltage of a power source lowers from the standard because of the load.

Washer
A flat ring used to form a seat for a bolt and nut.

Wavelength
The length of a single wave of light, radio transmission, etc.

Wedge
A type of robot that has a sloped front used for getting under the opponent and either flipping it or pushing it around.

Weld
To join pieces of metal by heat or pressure.

Wheel dresser
A tool for cleaning and straightening the face of a grinding wheel.

Wire tie
Also called a zip tie. A plastic or metal strip that is used to bind bundles of wires.

Work
Force acting through a distance.

Worm
The threaded cylinder or shaft designed to mesh with a worm gear.

Worm gear
A gear with helical teeth made to conform to the thread of the mating worm.

Yield strength
Amount of resistance to nonrecoverable or plastic deformation (permanent bending).

Yield stress
Highest stress before nonrecoverable or plastic deformation occurs.

Zircon
Fake diamond. In here only because I did not have another "Z" word.

INDEX

Note: Boldface numbers indicate illustrations.

About the Author

Chris Hannold has been involved in robotics for almost two decades. He is a well-known contributor to the BattleBot's™ Builder's Forum on the Internet where he sees hobbyists' questions on a daily basis. Mr. Hannold and his team have competed in BattleBots™, Robot Wars™, Robotica™, and local competitions. He also organizes and produces the NC Robot StreetFight (www.ncrsf.com) and is an active member of the East Coast Robot Alliance, South Eastern Combat Robotics, and the Society of Robotic Combat.

SOFTWARE AND INFORMATION LICENSE

The software and information on this diskette (collectively referred to as the "Product") are the property of The McGraw-Hill Companies, Inc. ("McGraw-Hill") and are protected by both United States copyright law and international copyright treaty provision. You must treat this Product just like a book, except that you may copy it into a computer to be used and you may make archival copies of the Products for the sole purpose of backing up our software and protecting your investment from loss.

By saying "just like a book," McGraw-Hill means, for example, that the Product may be used by any number of people and may be freely moved from one computer location to another, so long as there is no possibility of the Product (or any part of the Product) being used at one location or on one computer while it is being used at another. Just as a book cannot be read by two different people in two different places at the same time, neither can the Product be used by two different people in two different places at the same time (unless, of course, McGraw-Hill's rights are being violated).

McGraw-Hill reserves the right to alter or modify the contents of the Product at any time.

This agreement is effective until terminated. The Agreement will terminate automatically without notice if you fail to comply with any provisions of this Agreement. In the event of termination by reason of your breach, you will destroy or erase all copies of the Product installed on any computer system or made for backup purposes and shall expunge the Product from your data storage facilities.

LIMITED WARRANTY

McGraw-Hill warrants the physical diskette(s) enclosed herein to be free of defects in materials and workmanship for a period of sixty days from the purchase date. If McGraw-Hill receives written notification within the warranty period of defects in material or workmanship, and such notification is determined by McGraw-Hill to be correct, McGraw-Hill will replace the defective diskette(s). Send request to:

Customer Service
McGraw-Hill
Gahanna Industrial Park
860 Taylor Station Road
Blacklick, OH 43004-9615

The entire and exclusive liability and remedy for breach of this Limited Warranty shall be limited to replacement of defective diskette(s) and shall not include or extend to any claim for or right to cover any other damages, including but not limited to, loss of profit, data, or use of the software, or special, incidental, or consequential damages or other similar claims, even if McGraw-Hill has been specifically advised as to the possibility of such damages. In no event will McGraw-Hill's liability for any damages to you or any other person ever exceed the lower of suggested list price or actual price paid for the license to use the Product, regardless of any form of the claim.

THE McGRAW-HILL COMPANIES, INC. SPECIFICALLY DISCLAIMS ALL OTHER WARRANTIES, EXPRESS OR IMPLIED, INCLUDING BUT NOT LIMITED TO, ANY IMPLIED WARRANT OF MERCHANTABILITY OR FITNESS FOR A PARTICULAR PURPOSE. Specifically, McGraw-Hill makes no representation or warranty that the Product is fit for any particular purpose and any implied warranty of merchantability is limited to the sixty day duration of the Limited Warranty covering the physical diskette(s) only (and not the software or information) and is otherwise expressly and specifically disclaimed.

This Limited Warranty gives you specific legal rights, you may have others which may vary from state to state. Some states do not allow the exclusion of incidental or consequential damages, or the limitation on how long an implied warranty lasts, so some of the above may not apply to you.

This Agreement constitutes the entire agreement between the parties relating to use of the Product. The terms of any purchase order shall have no effect on the terms of this Agreement. Failure of McGraw-Hill to insist at any time on strict compliance with this Agreement shall not constitute a waiver of any rights under this Agreement. This Agreement shall be construed and governed in accordance with the laws of New York. If any provision of this Agreement is held to be contrary to law, that provision will be enforced to the maximum extent permissible and the remaining provisions will remain in force and effect.